P9-BJL-463

Transport systems in plants

Jeffrey Moorby

Agricultural Research Council
London

Longman London and New York

Longman Group Limited,
Longman House,
Burnt Mill, Harlow, Essex, U.K.

*Published in the United States of America
by Longman Inc., New York*

First published 1981

British Library Cataloguing in Publication Data

Moorby, Jeffrey
   Transport systems in plant. – (Integrated
   themes in biology).
   1. Plant physiology
   I. Title   II, Series
   581.4   QK881   80-41374

   ISBN 0-582-44379-2

Printed in Great Britain by William Clowes (Beccles) Ltd
Beccles and London.

Contents

Preface

Acknowledgements

1	**The evolution of transport systems in plants**	1
	1.1 Introduction	1
	1.2 Phloem	2
	1.3 Xylem	18
	1.4 Symplastic and apoplastic transport	28
2	**The movement of carbohydrates**	33
	2.1 Introduction	33
	2.2 Carbon assimilation	33
	2.3 Movement of carbohydrates into the phloem	35
	2.3.1 Movement in C_3 plants	35
	2.3.2 Movement in C_4 plants	39
	2.3.3 The kinetics of movement into the phloem	41
	2.4 The kinetics of movement through the phloem	47
	2.5 The possible, and sometimes impossible, mechanisms of phloem movement	60
	2.5.1 The pressure flow theory	63
	2.5.2 Active mechanisms	66
	2.6 Movement out of phloem	72
	2.7 Mathematical models of translocation	73

3 The movement of water and ions   76
  3.1 Introduction   76
  3.2 The movement of water into cells   76
  3.3 The movement of ions into cells   79
  3.4 The movement of water and ions into the xylem   82
  3.5 The movement of water and ions through the xylem   90
  3.6 The movement of water and ions into and through leaves   102
  3.7 The movement of ions in the phloem   104

4 Transport systems and plant growth   108
  4.1 Introduction   108
  4.2 The effect of environmental and other factors on transport systems   109
    4.2.1 Water   109
    4.2.2 Light   112
    4.2.3 Temperature   117
    4.2.4 Growth substances   120
  4.3 Transport systems and growth   121
    4.3.1 Transport and the growth of stem apices and leaves   123
    4.3.2 Transport and the growth of stems   139
    4.3.3 Transport and the growth of roots   141
    4.3.4 Transport and the growth of storage organs and fruit   142
      4.3.4.1 Transport and the growth of potato tubers   143
      4.3.4.2 Transport and the growth of cereal grains   146
      4.3.4.3 Transport and the growth of fruit   150
    4.3.5 Transport out of seeds and storage organs   155
  4.4 The movement of exotic substances   160

5 Conclusions   163

  Index   165

Preface

All plants, from unicellular aquatic algae to large terrestrial trees, are dependent for survival on the transport of a large variety of substances into, out of and through the plant. Obviously, the transport of materials within large plants is much more difficult than within a single cell. Hence, the evolutionary adaption of plants to permit the efficient functioning of the transport processes has resulted in many structural modifications. These modifications have been very successful because in very few instances do deficiencies in the internal transport processes appear to impose any limitations on plant growth.

I will be concerned initially with describing the tissues involved in transport in the different groups of plants and will try to show how these tissues have become more specialized. Some descriptions of developmental anatomy will be provided but discussions of how structure is related to the functioning of the tissues will be reserved for the later sections of the book. The second section will be concerned with the movement of carbon into simple and multicellular plants and its transport into and through the 'phloem' of the latter. The third section will consider the absorption and transport of ions and water.

The first three sections therefore, will, be concerned primarily with transport pathways and mechanisms. However any reasonably complete understanding must include some knowledge of the systems which supply material to the plant and so there will be some mention of transfer processes in the atmosphere and the soil. It will be impossible in the space available to discuss the operation of these in detail, or indeed, to provide a comprehensive coverage of the processes within plants. The aim will be to consider the fundamental problems involved and to discuss the various aspects in terms of the simplest and best understood situations. This discussion will then be extended to include other plant types and more unusual conditions.

It must be emphasized that although individual processes will be discussed separately, they operate as a complete system and that a plant functions as an integrated whole. The behaviour of a complex system such as a growing plant is often determined not by the individual parts of the system but by the interrelationships between the parts and how they respond to the constantly changing internal and external environment. The final part of the book will, therefore, extend the discussion beyond mechanistic considerations to the interrelationships between transport and other processes and growth.

J. Moorby
London, 1980

Acknowledgements

I would like to thank many friends, too numerous to list individually, for their help and encouragement whilst writing this book. I have spent many hours discussing with them the operation of plant transport systems. This book owes a lot to these discussions, and I apologise if I have misrepresented their ideas or twisted them to my own purpose.

It is, however, necessary to mention Professor M. J. Canny, Dr J. E. Dale, Professor D. S. Fensom, Dr J. Scobie, Dr R. G. Thompson and Dr E. J. Williams. We have worked together to develop techniques for the use of ^{11}C and have produced some of the unpublished data used in this book.

I am particularly grateful for the permission given by many authors and publishers to use tables and figures which have appeared elsewhere, to Mrs M. Carruthers and Miss K. Shah who have typed and retyped the manuscript and to Mrs C. Richardson and Mr M. Bone for help in producing some of the figures.

Finally, I have to thank my wife who has helped me sort out my ideas, corrected the English and spelling and helped in many other ways.

Chapter 1

The evolution of transport systems in plants

1.1 Introduction

In many plants water, carbohydrates and ions have to move over long distances and hence a variety of tissues and cell types have evolved to cope with this long-distance transport. The selective pressures which have resulted in these changes have been considerable and can be traced primarily to the action of two dominant evolutionary progressions; the development of multicellular forms and the transfer from aquatic to terrestrial habitats.

The tissues concerned with transport in all the plant groups are remarkably similar. I have chosen to emphasize these similarities rather than the differences. The two tissues of primary concern are the phloem and xylem. The general consensus of opinion is that most of the movement of sugars occurs in the phloem whereas ions move in both the phloem and xylem. Other materials such as organic forms of nitrogen and growth substances can also move in either tissue, but their transport is dependent on the species under consideration and is complicated by problems of synthesis and utilization. Water can move in both tissues. This is a gross oversimplification, as will be seen from the discussion which follows, but this is of no immediate concern. It is sufficient to be able to delimit the cells which need to be considered and the evidence which has made this possible has come mainly from two types of experiment. The first involved 'ringing' (removal of a strip of bark and phloem from the stem) or destroying these tissues by heat, coupled with chemical analyses. The second type of experiment involved the direct localization of the mobile materials by histochemical or autoradiographic techniques. Because of work of this type it is now agreed that long distance transport in the phloem takes place through the sieve elements, or their equivalents in lower plants, whereas the tracheids and vessels are the major conduits in the xylem.

Arguably the most primitive cells concerned with the transport of assimilates are to be found in the large brown algae, and these will be described in some detail. The types of sieve elements to be found in other plant groups will be described more briefly except for those of the angiosperms. This group has been used in most of the work on phloem transport and, possibly because of this, there is probably more dispute about the sieve cells in this group than in any other. Hence, the various interpretations of the structure and development of sieve cells in the angiosperms will be considered in more detail than for any other plant group.

For convenience, the development of the xylem will be discussed separately from that of the phloem. There is less dispute about the structure and functioning of xylem than of phloem, but again most of the discussion will be reserved for the xylem of the higher plants.

1.2 Phloem

In the algae there is a progression from unicellular to filamentous and colonial forms and eventually the development of large multicellular species such as the brown alga *Laminaria*. The distances over which the transport of assimilates must occur in species such as *Laminaria* are considerable, and autoradiography has shown that this transport takes place through specialised cells, the hyphae and trumpet hyphae (Fig. 1.1).

The trumpet hyphae are elongated cells which, under the light microscope, often appear to be narrower in the middle of the cell than adjacent to the transverse walls. This 'belling' of the ends of the cells is not always seen (cf. Fig. 1.2). It may be an artefact caused by the release of turgor pressure during specimen preparation or it may be caused by the stretching of the cells as the plant grows. The possibility of damage during sample preparation is a problem which bedevils all investigations of the structure of cells concerned with assimilate transport because they all seem to operate under a positive internal pressure.

There are small pores between adjacent hyphae and trumpet hyphae which are lined by the plasmamembrane (Fig. 1.3). These cells

Fig. 1.1 A typical trumpet hyphae in *Laminaria groenlandica* Rosenv. showing ▶ vacuoles, nucleus and plastids (arrowed) (after K. Schmitz and L. M. Srivastava, 1974).

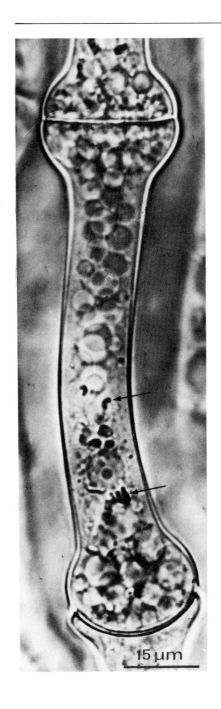

15 μm

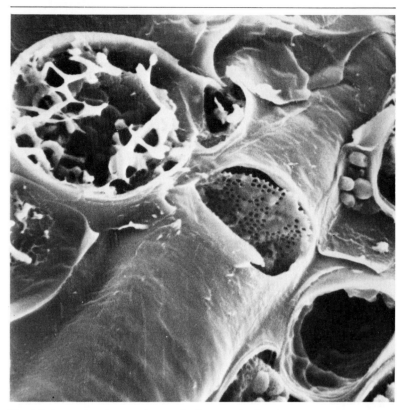

Fig. 1.2 A scanning electronmicrograph of trumpet hyphae in *Macrocystis pyrifera* (L) C. Ag. Part of the cell wall has been broken open to reveal the sieve plate between two cells which show little 'belling' (× 5400) (after J. H. Troughton and F. B. Sampson, 1973).

appear to form a reasonably efficient pathway for longitudinal transport through the plants, some tracer experiments suggesting that the speed of assimilate movement can be up to 0.8 m h^{-1}. When functional the trumpet hyphae appear to contain the usual organelles such as nucleus, mitochondria, plastids and endoplasmic reticulum. They also have many small vacuoles (Fig. 1.4). There do not appear to be associated cells analogous to the companion cells of the angiosperms. In older hyphae the protoplasmic connections through

Fig. 1.4 A cross-section of a sieve plate in *Alaria marginata* showing cytoplasmic ▶ connections between adjacent cells and endoplasmic reticulum near the pores (arrowed). NB. The white regions around the pores are probably preparation artefacts and not callose, cf. Fig. 1.5 (after K. Schmitz and L. M. Srivastava, 1975).

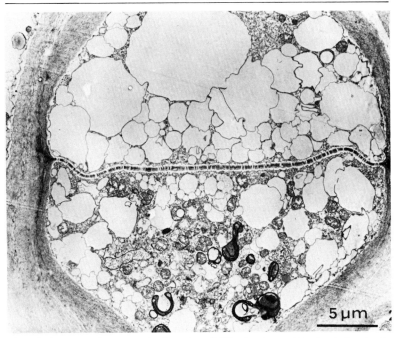

Fig. 1.3 A cross-section across the sieve plate dividing two trumpet hyphae in *Laminaria groenlandica* Rosenv. There are many vacuoles, plastids and mitochondria (after K. Schmitz and L. M. Srivastava, 1974).

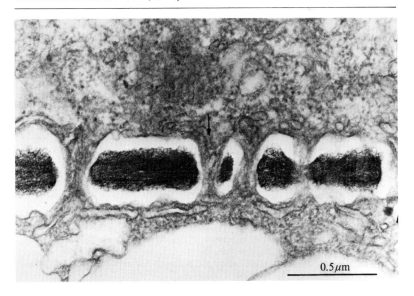

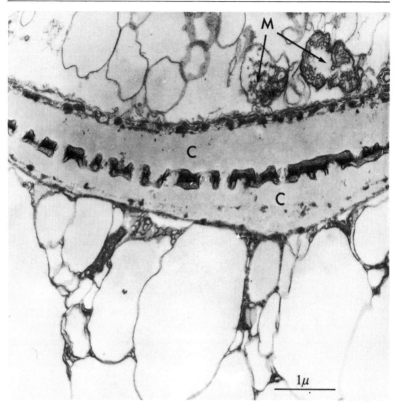

Fig. 1.5 A sieve plate in *Alaria marginata* which has been blocked with callose (C). NB. There are no cytoplasmic connections between the cells and the mitochondria (M) appear to be disintegrating (after K. Schmitz and L. M. Srivastava, 1975).

the pores are lost, the organelles disintegrate and the end walls become lined with callose (Fig. 1.5).

This general pattern of development and senescence can be found in all groups of plants but with variation in detail. Because all the cells concerned with assimilate transport have similar groups of pores, the sieve areas, there is now a tendency to refer to them all as sieve cells or sieve elements. The former term is usually applied to those cells which have sieve areas which are equally well differentiated in the lateral and

Fig. 1.6 The plasmodesmatal connections (PD) between two young leptoids (LEP) of ▶ *Polytrichum commune* L. The plasmalemma (PE) can be seen on both sides of the cell wall and there is a dictyosome (D) and some small vacuoles in the cell on the left (after C. Hébant, 1970).

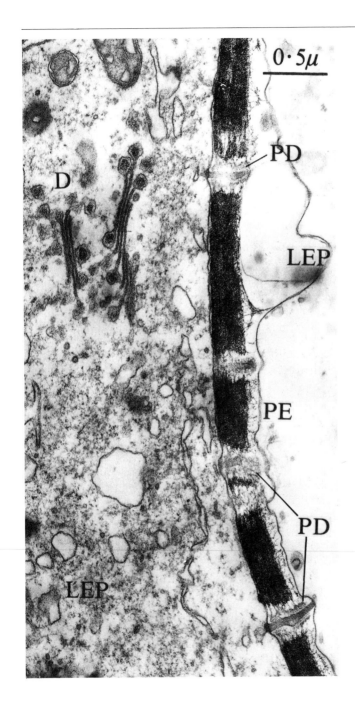

end walls. If some of the sieve areas, say those on the end walls, are more highly differentiated than those on the other walls the cell is usually called a sieve element. In all plant groups the cells are long and thin and are often short-lived. In perennial monocotyledonous plants such as palms, however, it has been estimated that some sieve cells can function for up to 100 years. It is now generally accepted that functional sieve elements have a positive internal pressure and retain some measure of cytoplasmic organization, but the fine details of the organization are hotly contested. In contrast, the xylem cells, which are concerned primarily with the transport of water and ions, are dead, open conduits which operate at negative internal pressures.

The largest and most prominent part of a moss is the gametophyte and this contains two types of cells adapted to transport, the sieve cells, or leptoids, which have been shown by autoradiography to transport ^{14}C-labelled assimilates, and the hydroids which seem to be concerned with the conduction of water. The transverse walls of the leptoids have pores which seem to be enlarged plasmodesmata and are similar in size to the pores found in the sieve elements of the protophloem of some vascular plants (Fig. 1.6). As in these latter cells, the nucleus of leptoids degenerates during differentiation but some endoplasmic reticulum persists and is often associated with the pores. A further point of similarity with higher plants is the association between the leptoids and enzyme-rich parenchyma cells.

In ferns the tissues concerned with the long distance transport of assimilates are the sieve cells and phloem parenchyma. They are very similar to the phloem found in angiosperms whereas the equivalent tissues in some of the more primitive vascular plants show less highly developed characteristics. For example, in *Isoetes*, the end walls of the sieve cells have pores but the lateral walls have only plasmodesmata. In the closely related *Lycopodium* and *Selaginella*, however, all the plasmodesmata develop into pores on both the lateral and end walls. During the differentiation of the sieve cells the nucleus degenerates, the extent of the degeneration increasing until it is complete in the higher ferns. The endoplasmic reticulum becomes dispersed but filamentous structures can be found in mature cells and in the pores connecting these cells. These filaments are not, however, considered to be the same as the p-protein (phloem protein) found in angiosperms. Structures common in the sieve cells of ferns are small, highly refractive spherules, probably of protein, surrounded by a unit membrane; their function is unknown. In some ferns there is some thickening of the cell walls of the sieve cells to form a nacreous wall the name arising from the lustrous appearance under the light microscope.

The parenchyma cells associated with the sieve cells have dense protoplasts with many mitochondria and ribosomes. In *Lycopodium* the parenchyma cells are connected to adjacent sieve cells by

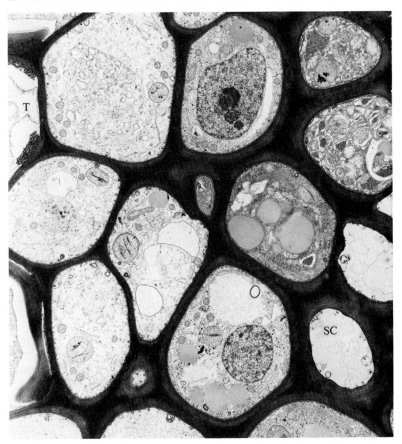

Fig. 1.7 A cross-section of the vascular system of *Lycopodium lucidulum* Michx. with a tracheid (T) on the left and sieve cell (SC) on the right. The parenchyma cells nearest the sieve cell have denser cytoplasm and thicker walls than those nearest the tracheid ($\times$ 7300) (after R. D. Warmbrodt and R. F. Evert, 1974).

numerous plasmodesmata. In contrast, there are no plasmodesmatal connections between the tracheids and adjoining parenchyma cells and the latter have much less dense protoplasts than those near the sieve cells (Fig. 1.7).

The sieve cells of the gymnosperms are generally very similar to those of the angiosperms. They have pores in both the lateral and end walls and often form distinct nacreous walls. These latter are secondary walls, that is they are laid-down after the surface area of the cells ceases to increase, but they contain no lignin and are absent from the sieve areas. The major point of difference of the sieve cells of gymnosperms from those of the angiosperms is that they have no p-protein. Phloem-protein was usually called slime until the mid-1960s. It can occur in several distinct morphological forms but is

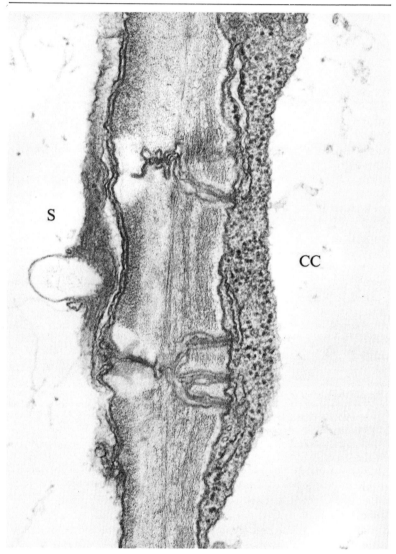

Fig. 1.8 Plasmodesmata between a sieve cell (S) and companion cell (CC) in *Nicotiana tabaccum*. In each plasmodesmata there is a single cavity leading to the sieve cell which branches on the side of the companion cell (× 60 000) (after K. Esau, 1969).

not endoplasmic reticulum. In contrast, the filamentous structures which can be found in the sieve cells of gymnosperms, and which pass from cell to cell through the sieve pores, are probably endoplasmic reticulum or derived from endoplasmic reticulum. Associated with the sieve cells are albuminous cells which develop from ray cells and seem to be analogous to the companion cells of the angiosperms. They have dense cytoplasm and are connected to the sieve cells by numerous plasmodesmata which branch on the side of the albuminous cells.

A sieve cell/companion cell pair in the angiosperms is formed by the unequal division of a procambial initial. The daughter cells are connected by numerous plasmodesmata which often branch on the side of the companion cell. In contrast, the plasmodesmata passing from one sieve cell to another sieve cell or to a parenchyma cell are not usually branched (Fig. 1.8). The companion cells retain their nucleus and have dense cytoplasm containing many mitochondria, ribosomes and plastids and much endoplasmic reticulum. As development of the sieve cell proceeds endoplasmic reticulum appears to lie in the cytoplasm alongside the plasmodesmata and there is some vacuolation. The nuclei, and some of the other organelles, start to break down but some mitochondria and plastids can usually be found throughout the life of the sieve cells. The plastids usually have fewer thylakoid membranes but can accumulate starch or protein.

The cytoplasm of the sieve cells is rich in protein. This p-protein has been found in the sieve cells of all the dicotyledonous, and in many, but not all, the monocotyledonous species examined. It is, for example, not found in *Zea mays*, but in this species a network of endoplasmic reticulum persists. There appears to be an early association between the p-protein and the endoplasmic reticulum. This may have some connection with the assembly of the p-protein structures, which may be tubular, fibrillar or granular, but the lack of ribosomes at this stage of development would suggest that there was no connection with protein sythesis.

As this production of p-protein proceeds the plasmodesmatal pores start to widen and form the sieve pores. The formation of sieve pores often seems to be accompanied by the production of callose platelets on the walls of the sieve plate around, and sometimes lining, the sieve pores. Until the early 1960s it was accepted that the sieve cells were conduits connected by open sieve pores which contained little to impede the flow of solution. The appearance of callose and 'slime' in some preparations was, therefore, thought to result from damage during preparation of the sections or the senescence or death of the sieve cells.

This view was challenged when some of the first electron microscope studies showed the prescence of material plugging the sieve pores (cf. Fig. 1.9). In addition, a return was made to the study of

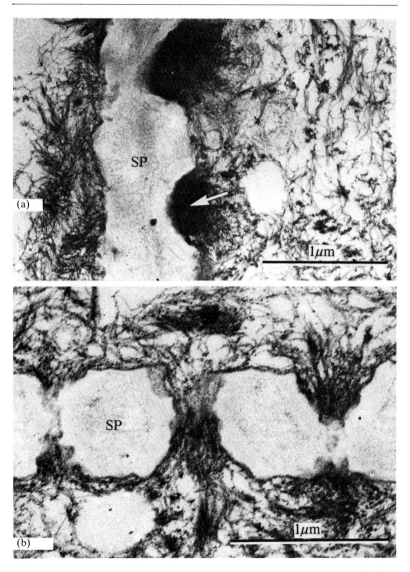

Fig. 1.9 Displacement of the contents of sieve tubes during sample preparation. A vascular bundle was dissected from a celery (*Apium graveolens* L.) petiole and a section 3.5 cm long was fixed in gluteraldehyde and divided into seven 0.5 cm pieces. (*a*) was within 1 cm of the end of the 3.5 cm section and (*b*) in the middle. The sieve tube contents accumulated on the right hand side of the sieve plate (SP) in (*a*), i.e. the side nearest the end of the section. There was no similar 'piling-up' in (*b*) (after G. P. Dempsey, S. Bullivant and R. L. Bieleski, 1975).

sections of living phloem and it was claimed that strands of material passed across the cells and from cell to cell through the sieve pores. It was also claimed that the strands were bounded by membranes and that the contents of the strands moved from cell to cell. The speed of movement was 3–5 cm h^{-1} and it was sometimes possible to see movement in opposite directions in the same sieve cells.

There is still no general agreement on the existence of the membrane-bound strands. However it is now generally agreed that structures can be found in sieve cells; often composed of p-protein. The arrangement of the p-protein within the cell is still disputed. The various interpretations depend on the methods of specimen preparation and the beliefs of the workers concerning the types of structure which can be reconciled with physiological information on the transport processes. Certainly, different methods of sample preparation can produce very different images, and at the present time it is impossible to know which, if any, are near the truth.

The aim of any method of sample fixation and preparation is that the structure of the functioning cell is preserved with the minimum of disturbance. The 'nicest', and probably most easily understood micrographs are probably produced by the traditional chemical fixatives such as gluteraldehyde and acrolein (cf. Fig. 1.5). Nevertheless it can, and has been, argued that since these take several seconds, or minutes, to fix the tissue there is a strong possibility that artefacts are produced during fixation. For example, if the sieve cells are 2.5×10^{-4} m in length and the speed of translocation is 2.78×10^{-5} ms^{-1} (100 cm h^{-1}) there could be movement through more than one sieve cell before fixation occurred. This would almost inevitably cause distortion of the precipitating structures.

In contrast, it is possible under suitable conditions to freeze intact, functioning, sieve cells in approximately 0.02 s. In this time a particle in the translocation stream would have moved only 0.6×10^{-6} m; a fraction of the length of a sieve cell. Under these conditions, therefore, damage might be expected to be less than when using a chemical fixative. Damage might arise, however, if the rate of freezing is insufficiently fast to prevent the formation of ice crystals within the cells and Brownian motion during freezing could also cause disruption. The frozen material can be processed by freeze substitution or freeze fracturing and etching. In the former, the frozen material is infiltrated with acetone or propylene at -70 °C to prevent movement and loss of water-soluble solutes. It can then be embedded and sectioned using standard methods (Fig. 1.10). Alternatively, the frozen tissue can be fractured in the region of the phloem, shadowed with carbon and platinum and replicated before viewing in the electron microscope (Fig. 1.11). The micrograph in Fig. 1.10 differs from some of the others and Fig. 1.11, in showing the sieve pores almost completely free of

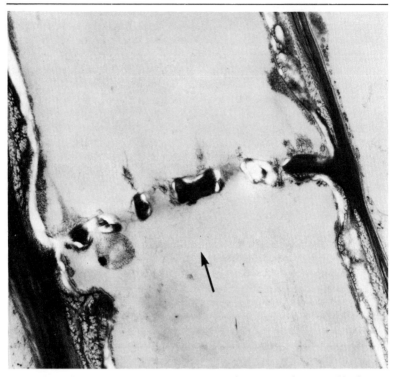

Fig. 1.10 Section through a sieve plate of soybean (*Glycine max*) prepared by freeze substitution. The arrow indicates the probable direction of flow. There is some parietal endoplasmic reticulum and small strands near the pores. The pores appear to be quite open (× 19 500) (after D. Fisher, 1975).

material. There is some pariatal endoplasmic reticulum and a few small strands near the pores. Figure 1.11*a* and *b* are more difficult to interpret because they are in relief. In both, however, filamentous material can be seen in the pores. The quality of the preservation can be seen in Fig. 1.11*a* where the tonoplast can be easily distinguished in the adjacent phloem parenchyma cell.

Several problems are common to all methods of examination. One is the avoidance of surge effects produced by the release of pressure in

Fig. 1.11 (*a*) Section through a sieve pore of a petiole of *Nymphoides peltata* prepared ▶ by freeze etching and (*b*) a single pore adjacent to the cell wall at a higher magnification. ER is endoplasmic reticulum in the adjacent parenchyma cell; T is the tonoplast; V the vacuole; W is the cell wall; P the plasmalemma of the sieve cell; Z frozen-out solutes; F filaments in the sieve pore. The arrow indicates the direction of shadowing. The scale is 1 μm (after R. P. C. Johnson, 1973).

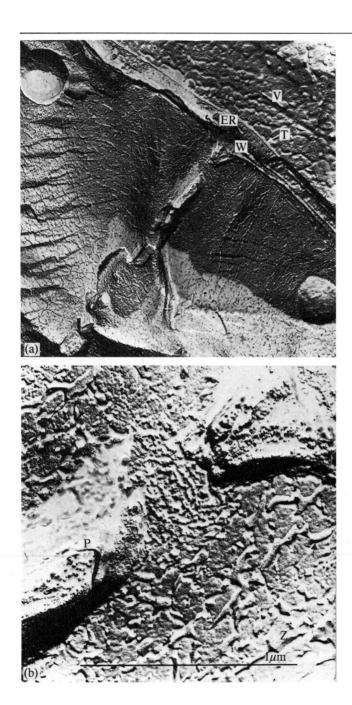

the sieve tubes. This can be a major problem in unfrozen material, especially near the ends of segments removed for fixation (cf. Fig. 1.9*a* and *b*). Associated with this problem is that of callose formation. Callose is a $\beta - 1, 3$ glucan which can be produced very rapidly in sieve tubes and which can block the sieve pores (Fig. 1.12). There is some dispute about the extent to which callose is present in undisturbed functioning sieve cells. In many species callose is produced in senescent sieve tubes and in these circumstances can effectively prevent any further transport. In some perennial species, however, callose may be deposited as the plant becomes dormant in the autumn and

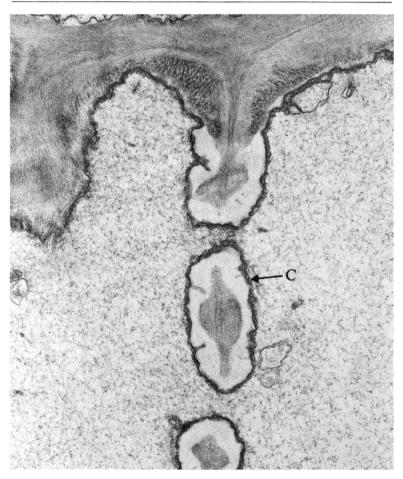

Fig. 1.12 A mature sieve plate of *Cucurbita maxima* showing a deposition of callose (C) on the sieve plate ($\times$ 21 700) (after K. Esau, 1969).

16

disappears on the resumption of growth the following spring. It is always difficult to be sure that translocation was actually taking place through the cells examined. In some instances this can be done by combined autoradiography and electron microscopy. An alternative method was used in the production of Fig. 1.11a. This was obtained from a *Nymphoides* petiole which had been dissected to leave only the central vascular bundle. The leaf was then exposed to $^{14}CO_2$ and the

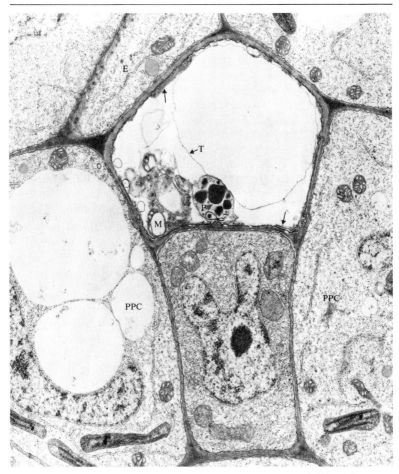

Fig. 1.13 Sieve cell and companion cell of *Lemna minor* surrounded by phloem parenchyma (PPC) and endodermis, E. The sieve cell has less dense cytoplasm than the other cells but does contain endoplasmic reticulum, arrowed, a mitochondrion, M; a plastid, P; and a tonoplast, T. The material was fixed in gluteraldehyde-paraformaldehyde ($\times$ 20 700) (after J. E. Melaragno and M. A. Walsh, 1976).

material only taken for sectioning if translocation had taken place through the dissected region.

It must be obvious from this discussion that there is no agreed interpretation of sieve tube structure. Opinions vary depending on the techniques used for sample preparation and the views of the protagonists concerning the possible mechanisms of movement through the sieve tubes. Few things can be stated with any degree of certainty. The plasmalemma seems to remain intact throughout the life of a sieve cell and to line the sieve pores. In contrast, until recently the tonoplast was thought to be lost in the early stages of development (cf. Fig. 1.11a). In *Lemna minor*, however, the tonoplast can be found in mature sieve cells (Fig. 1.13). In the sieve cells of angiosperms there is an appreciable amount of p-protein which can be found as crystalloid structures or strands, the latter sometimes banded or possibly helical in structure. It seems reasonable to assume that the more highly organised types of structures that are found are nearest to what actually exists in the functioning sieve cells. (According to this interpretation, the 'slime' often found in early work, or the more amorphous p-protein in some contemporary work is probably the result of bad fixation and handling in specimen preparation.) Since the suggestions concerning the nature and distribution of the filaments or strands are so closely linked function they will not be pursued here, but will be reserved for the discussion of the various theories concerning the mechanism of movement through the sieve cells.

1.3 Xylem

The major environmental pressure during the evolution of terrestrial plants must have been the necessity to obtain and conserve water. We see, therefore, the development of rhizoids, which in the liverworts are unicellular, but become multicellular, and more extensive in the mosses. The clubmosses and ferns developed roots, as did all the higher plants. Some of the ferns were the first species to develop root hairs.

The exposed tissues of land plants are very susceptible to dessication, but the cuticle, which helps to prevent this dessication, has the disadvantage of also inhibiting the uptake of carbon dioxide. The first adaptation to aid the exchange of gases between the plant and the atmosphere was the development in liverworts of barrel pores. These have a similar function to the stomata of higher plants, but their operation is different. True stomata are found in the capsules of some mosses, but they are usually absent from the leaves where diffusional problems are small because moss leaves are often only one cell thick. The ferns and higher vascular plants have stomata on the leaves. In some fern species the stomata are depressed below the surface of the leaf as a further aid to conserve water. This type of adaptation is seen at its most extreme in the xerophytic gymosperms and angiosperms.

The exposed tissues of terrestrial plants would die from dessication if the water evaporated was not replaced. This replacement is dependent on the existence of an efficient transport pathway between the sites of absorption and evaporation. In the gametophytes of mosses this pathway is provided by the hydroids; an axial strand of thin-walled cells (Fig. 1.14). During their development the cytoplasm disappears. There seems to be no conclusive evidence of any lignified secondary thickening of the cell walls, and in this regard the hydroids resemble the unthickened tracheids in the protoxylem of higher plants.

The greater size and complexity of the clubmosses and ferns have led to the production of a more complicated vascular system than that found in the mosses. They are the most primitive plants to develop xylem and fossil arborescent forms, and a few living species, produce secondary vascular tissue. The conducting cells of the xylem are tracheids; dead lignified cells with pitted walls. A pit is a region of a cell wall where no thickening is laid down on the primary cell wall. The pits in adjacent cells are usually contiguous, forming a pit pair, the members of which are separated by a boundary composed of the middle lamella and the two primary cell walls. This boundary is often perforated by plasmodesmatal connections. Pits are often clumped together to form a pit field (Fig. 1.15). In most fern species the pits are of the simple type described, but in *Ophioglossum* there are more complicated bordered pits (see p. 22). Although the pits offer less resistance to cell transport than does the remainder of the cell wall, there is still a major barrier. An evolutionary trend which can be followed, especially in higher plants, is the removal of this barrier and the formation of interconnecting tubes from files of cells; the vessels. These structures can be found in some of the higher ferns, e.g. *Pteridium*.

In addition to these evolutionary trends at the cellular level, it is also possible to follow trends at the tissue level. *Gleichenia*, for example, has a simple cylindrical vascular system, a protostele, containing an inner mass of xylem surrounded by phloem. In contrast, *Dryopteris* has a much more complex vascular system, a dictyostele, composed of many small anastomosing vascular bundles each of which often resemble a protostele. In some species, however, the vascular bundles do not have concentric phloem but more nearly resemble the bicollateral bundles found in the cucurbits, with internal and external phloem separated by the xylem.

As in the ferns the conducting cells in the xylem of gymnosperms are almost entirely tracheids and vessels are formed in a few species. The tracheids usually have a diameter of 20–50 μm and are typically about

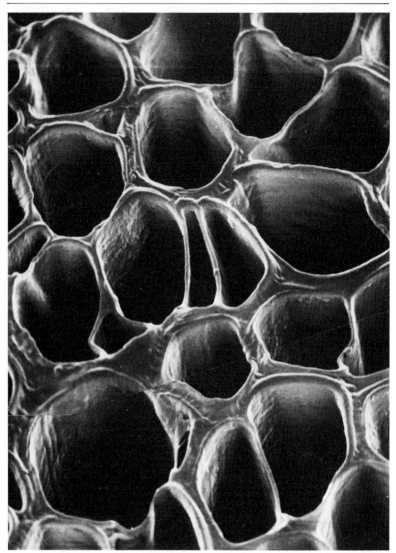

Fig. 1.14 Hydroids in the central vascular strand of the stem of *Polytrichum commune* L. There appears to be no sculptured secondary thickening and the cell walls are smooth (× 5200) (after C. Hebant, 1974).

100 times longer. Secondary xylem and phloem are formed from a cambium. In some, the ring-porous species, the tracheids formed during the early part of the season have a larger cross-sectional area,

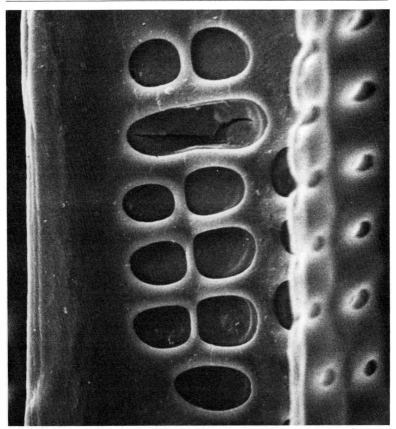

Fig. 1.15 A group of pits viewed from inside a vessel in the xylem of *Nothofagus fusca* Hook f. The pit membranes are intact except that in the large oblong pit which has been broken during sample preparation (× 3400) (after B. A. Meyland, and B. G. Butterfield, 1971).

and thinner walls, than those formed later. In contrast, in the diffuse-porous species there is little variation in the diameter of the tracheids throughout the season.

Transport between the tracheids is facilitated by the presence of many bordered pits. For example, there can be up to 90 in the radial walls of tracheids in the early wood of *Pseudotsuga*, but only about 10 per tracheid in the late wood. The bordered pits are more complex than the simple pits described on page 19 and can act as valves. The edges of both sides of the pit arch over the membrane separating the cells to form the pit borders (Fig. 1.16*a*). This structure is circular in many species and typical dimensions of bordered pits in *Pinus sylvestris* are 10 to 20 μm for the outer diameter of the pit and 3 to 5 μm for the pore through the border (Fig. 1.16*b*). In other species, the structure takes the form of a slit with the long axes in the two cells often at an angle to each other (Fig. 1.16*c*). There is a thickened area in the

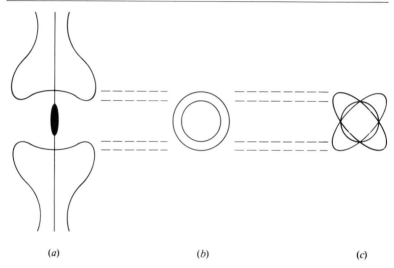

(*a*)   (*b*)   (*c*)

Fig. 1.16 The structure of bordered pits (*a*) A cross-section through a pit showing the secondary wall arching over the aperture. The dividing membrane is composed of a thickened central area, the torus, surrounded by the margo. (*b*) and (*c*) are surface views, (*b*) a circular pit and (*c*) a pit with elliptical openings at an angle to each other.

centre of the membrane separating the cells, the torus, and this is supported by radial strands in the surrounding region, the margo (Fig. 1.17). The exact nature of the margo is still in dispute. Some workers suggest the presence of an actual membrane, whereas others believe there are openings between the radial strands. Certainly, lateral transport is relatively easy and small particles have been shown to pass readily between tracheids in several conifers.

Any excess pressure on one side of a bordered pit will tend to move the torus towards the side of least pressure. In addition, if air enters a tracheid, surface tension forces generated by the evaporation of the liquid between the pit membrane and the pit border can also cause the movement of the membrane so tending to aspirate, or seal the pit (Fig. 1.18). The extent of any sealing will be the resultant of the forces causing the pit membrane to move and of the tension developed in the radial strands of the margo. Calculations have suggested that when the large pits of the early wood of *Pinus sylvestris* are aspirated there must

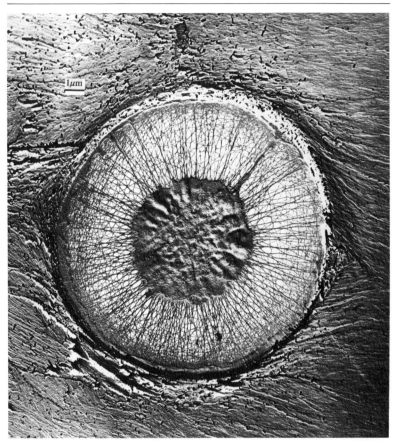

Fig. 1.17 A section showing the membrane in a bordered pit of *Abeis grandis*. The torus can be seen in the centre of the membrane. The surrounding margo has thick radial strands and with an interconnecting network of thinner strands (after J. A. Petty, 1972).

be some creep in cellulose fibres of the radial strands of the margo. This movement can be reversible, and the pits can reopen, unless there are permanent changes in the pit membrane; probably the formation of hydrogen bonds between the membrane and the border of the pit.

In addition to serving as a transport pathway the tracheids provide structural support for the plant and their rigidity is increased by the secondary thickening that is laid down. In the protoxylem, formed in the young extending parts of the plant, this thickening is usually helical or annular and continues to provide support to the cell wall as the latter is being stretched. In the later-formed meta- and secondary xylem the

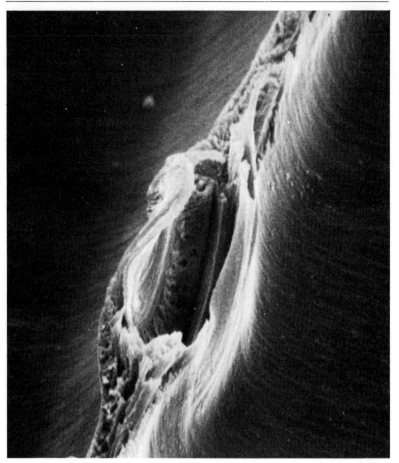

Fig. 1.18 An aspirated bordered pit in *Pinus*. The torus has been forced against the left side of the pit aperture and would restrict the lateral movement through the pit (× 6100) (after B. A. Meylan and B. G. Butterfield, 1971).

thickening forms a rigid network over the inside of the cell. In perennial species the cells actively concerned with conduction are often restricted to the current annual ring or a few recently formed rings of xylem. The older layers of xylem are gradually converted to heart wood by the deposition of substances such as resins and tannins and the formation of tyloses, that is, the invasion of vessels by the protoplasts of nearby parenchyma cells.

The progression in the types of thickening found in the gymnosperms is also found in the angiosperms (Fig. 1.19 and 1.20). A

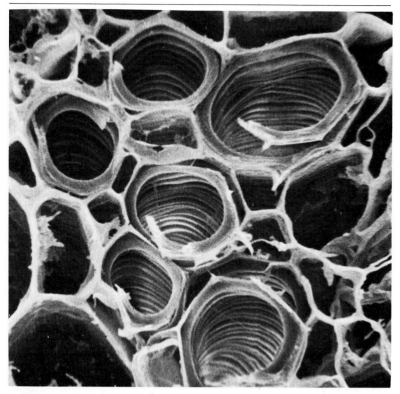

Fig. 1.19 Protoxylem vessels of *Gossypium hirsutum* L. showing helical thickening some of which has been detached from the cell wall during sample preparation (× 1700) (after J. H. Troughton, and F. B. Sampson, 1973).

major difference from the gymnosperms is the widespread occurrence of xylem vessels. These latter are produced by the dissolution of the end walls of adjoining cells to produce very long and wide channels for water and ion movement (Fig. 1.21).

It can be seen that the xylem is a tissue well adapted to allow the free movement of solutions throughout the plant. Because of its orientation it is particularly well suited to transport along the axis of the plant but the numerous pits in the radial and tangential walls of the cells also allow free lateral movement where necessary. Although transport can occur through much of the xylem it does not necessarily follow that this happens. When only small amounts of solutions are moving this usually takes place through the wider vessels and tracheids, i.e. those with the least resistance to flow. As the amount moving increases more and smaller cells become involved.

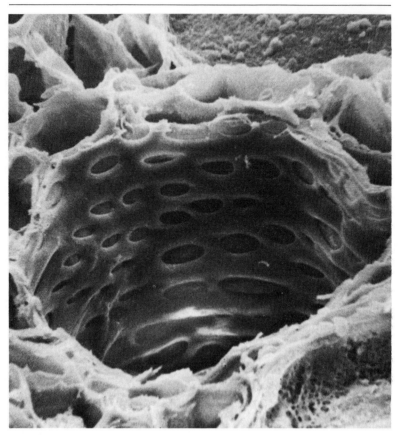

Fig. 1.20 A xylem vessel from the stem of *Cucumis sativus* showing reticulate thickening (× 1200) (after J. H. Troughton and L. A. Donaldson, 1972).

Fig. 1.21 The remnents of end walls between vessel members in the xylem of (*a*) ▶ *Knightia excelea* R. Br (× 1700) and (*b*) *Ulmus* (× 1300). In (*a*) the remaining cell wall forms a rim which is much less pronounced in (*b*). The pit membranes can be seen in the section of *Knightia* where the lateral wall was sectioned through a row of pits (after B. A. Meylan and B. G. Butterfield, 1971).

26

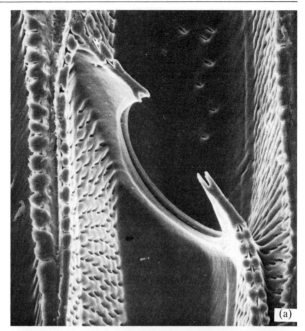

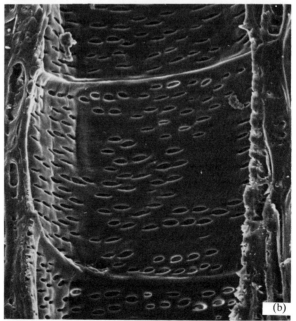

1.4 Symplastic and apoplastic transport

It is possible that the most fundamental difference between the xylem and the phloem is that the former tissue is mainly composed of dead cells whereas the functioning cells of the phloem are living. As such, the phloem cells have a plasmalemma which lines the sieve pores and hence is continuous between adjacent cells. In addition, the plasmalemma can be found lining the plasmodesmata which connect the sieve cells to adjacent living cells. The plasmodesmata, with an internal diameter of 30–60 nm, may simply provide a fluid connection between cells, but electromicrographs often show some electron dense material in the plasmodesmatal pore. This has been interpreted as forming an inner desmotubule the inner and outer diameters of which are usually in the ranges 7–10 and 16–20 nm respectively. The desmotubule is connected to the endoplasmic reticulum in the cells on either side of the plasmodema and often contains a central rod (Fig. 1.22). It is not clear to what extent the desmotubule and the space between the desmotubule and the plasmalemma form independent channels for movement between cells, nor whether there is an open annulus outside the desmotubule at the necks at each end of the plasmodesma. Nevertheless, even though they rarely occupy more than 1 per cent of the surface of a cell, it appears likely that the area available for transport is usually sufficient to provide a more efficient means of transport between two cells than movement across two plasmalemmata and the intervening cell wall. The rate of movement through the plasmodesmata is dependent to some extent on the rates of cyclosis within the cells since these will affect the concentration differences between the ends of the pores. Other modifying influences will be the dimensions of the plasmodesmata and the extent to which the channels between the cells are occluded.

It should be noted that the plasmalemma and the cytoplasm within the plasmalemma are continuous from cell to cell. This continuum within the plasmalemma has been called the symplast. The non-symplastic part of the plant, the apoplast, comprises the dead tissues of the xylem and those parts of the plant which lie outside the plasmalemma such as the cell walls.

It is possible, therefore, to envisage two transport pathways, one outside and the other inside the plasmalemma. The operation of these two systems can be detected in some circumstances and will be illustrated in later sections. Because interchange between the two pathways involves passage across a membrane the possibility arises of some interference to, or control of, the interchange at the plasmalemma. The importance of this interchange between symplast and apoplast is highlighted by the development of specialized cells in

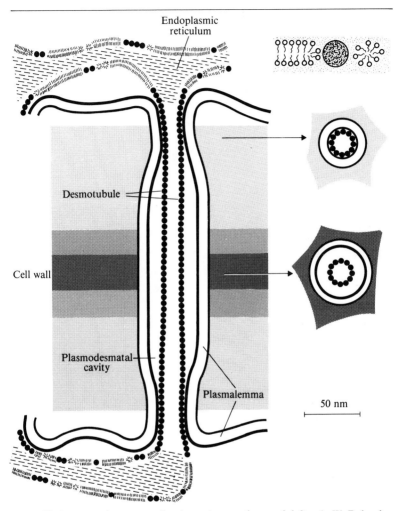

Fig. 1.22 A proposed structure of a plasmodesmata (see text) (after A. W. Robards, (Ed) 1971).

regions where this movement is pronounced. These are the transfer cells.

The onset of transfer cell differentiation appears to coincide with the start of very active transport in the region in question; for example leaves which are starting to export assimilates and the cotyledons of germinating seeds. The cells start to form ingrowths of the walls which remain covered with the plasmalemma. The resulting increase in

29

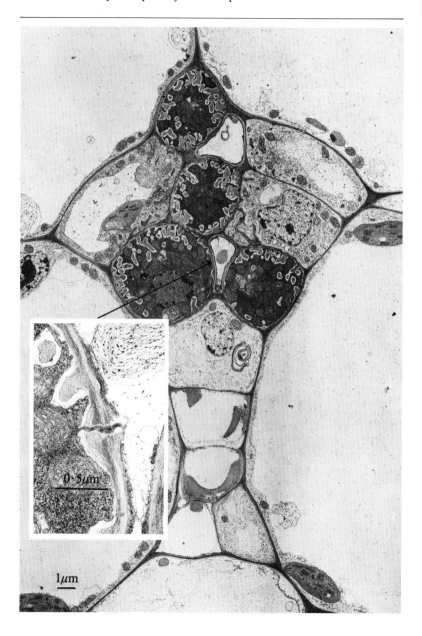

0·5μm

1μm

surface area provides, therefore, a greater area for absorption into, or passage from the cell (Fig. 1.23). The ways in which the transfer cells are involved in the functioning of the transport systems of the plant will be described in later sections.

Further reading and references

ARONOFF, S., DAINTY, J., GORHAM, P. R., SRIVASTAVA, L. M. and SWANSON, C. A. (1975) *Phloem Transport*. Plenum Press, New York.

DEMPSEY, G. P., BULLIVANT, S. and BIELESKI, R. L. (1975) The distribution of p-protein in mature sieve elements of celery, *Planta (Berl.)*, **126**, 45–59.

ESAU, K. (1969) *The Phloem, Encyclopedia of Plant Anatomy*, Vol. 2. Gebruder Borntraeger, Berlin.

FISHER, D. (1975) Structure of functional soybean sieve elements, *Plant Physiol.*, **56**, 555–60.

GUNNING, B. E. S. (1977) Transfer cells and their role in transport of solutes in plants. *Sci. Prog.*, Oxf., **64**, 539–68.

GUNNING, B. E. S. and PATE, J. S. (1969) 'Transfer cells' plant cells with wall ingrowths. Specialised in relation to short distance transport of solutes – their occurrence, structure and development, *Protoplasma*, **68**, 107–33.

HÉBANT, C. (1970) Histologie Végetale – Aspects infrastructurant observes au cours de la differenciation du phloeme (leptome) dans la tige feuillee de quelques Mousses Polytrichales, *C. R. Acad. Sc. Paris. Ser. D*, **271**, 1361–3.

HÉBANT, C. (1974) Studies on the development of the conducting tissue-system in the gametophytes of some polytrichales. II. Development and structure at maturity of the hydroids of the central strand, *Journ. Hattori Bot. Lab.* No. 38, 565–607.

JOHNSON, R. P. C. (1968) Microfilaments in pores between frozen-etched sieve elements. *Planta (Berl.)*, **81**, 314–32.

JOHNSON, R. P. C. (1973) Filaments but no membranous transcellular strands in sieve pores in freeze-etched, translocating phloem, *Nature*, **244**, 464–6.

JOHNSON, R. P. C. (1978) The microscopy of p-protein filaments in freeze-etched sieve pores, *planta (Berl.)*, **143**, 191–205

MELARAGNO, J. E. and WALSH, M. A. (1976) Ultrastructural features of developing sieve elements in *Lemna minor* L. – the protoplast. *Amer. J. Bot.*, **63**, 1145–57.

MEYLAN, B. A. and BUTTERFIELD, B. G. (1971) *The Three-Dimensional Structure of Wood*. Chapman and Hall, London.

PARTHASARATHY, M. V. (1975) Sieve-element structure, pp. 3–38, in Zimmermann, M. H. and Milburn, J. A. (eds), *Transport in Plants I: Phloem Transport, Encyclopedia of Plant Physiology*, Vol. 1 (New Series) Springer-Verlag. Berlin.

◀ **Fig. 1.23** A cross-section of a minor vein of groundsel (*Senecio vulgaris*). The two small relatively empty cells near the top of the vein are sieve tubes. The four cells with dense cytoplasm and obvious wall ingrowths are companion cells which have developed into one type of transfer cell. The three cells with less dense cytoplasm near the top of the vein are a phloem parenchyma cell which have differentiated into another type of transfer cell. These also have some wall ingrowths. The insert shows a plasmodesma connecting the sieve tube and companion cell. This is from a serial section and the line between the main picture and insert connects corresponding positions on the cell wall. (Main picture × 6200, insert × 36 400) (after B. E. S. Gunning, 1977).

The evolution of transport systems in plants

PARTHASARATHY, M. V. (1975) Sieve-element structure, pp. 3–38, in Zimmermann, M. H. and Milburn, J. A. (eds), *Transport in Plants I: Phloem Transport, Encyclopedia of Plant Physiology*, Vol. 1 (New Series) Springer-Verlag. Berlin.

PETTY, J. A. (1972) The aspiration of bordered pits in conifer wood, *Proc. Roy. Soc. Ser. B.*, **181**, 393–400.

RAVEN, J. A. (1972). The evolution of vascular land plants in relation to supracellular transport processes. Advances in Botanical Research Vol. 5, pp 153–219, Ed. Woolhouse, H. W., Academic Press, London.

ROBARDS, A. W. (1971) *Dynamic Aspects of Plant Ultra Structure*. McGraw Hill, London.

SCHMITZ, K. and SRIVASTAVA, L. M. (1974). Fine structure and development of sieve tubes in *Laminaria groenlandica* Rosenv, *Cytobiol*, **10**, 66–87.

SCHMITZ, K. and SRIVASTAVA, L. M. (1975) On the fine structure of sieve tubes and the physiology of assimilate transport in *Alaria marginata*, *Can. J. Bot.*, **53**, 861–76.

TROUGHTON, J. H. and DONALDSON, L. A. (1972) *Probing Plant Structure*. A. H. & A. W. Read, Wellington.

TROUGHTON, J. H. and SAMPSON, F. B. (1973) *Plants — A Scanning Electron Microscope Survey*. John Wiley & Sons, Australia Pty Ltd., Sydney.

WARMBRODT, R. D. and EVERT, R. F. (1974) Structure of the vascular parenchyma in the stem of *Lycopodium lucidulum*. *Amer. J. Bot.*, **61**, 437–43.

Chapter 2

The movement of carbohydrates

2.1 Introduction

The way in which the carbon enters the plant and is assimilated affects
the pathway by which it enters the phloem. Hence, this chapter starts
with a brief description of the different types of carbon fixation. This
will be followed by an account of the movement of carbon into,
through, and out of the phloem.

2.2 Carbon assimilation

The carbon fixed by the plant is absorbed from either the atmosphere
or surrounding solution. The atmosphere contains approximately 0.03
per cent carbon dioxide and a solution which is in equilibrium with the
surrounding atmosphere will also contain about the same concentration plus the carbonate and bicarbonate ions which are in equilibrium.
The absolute and relative concentrations will depend on, for example,
the rate of stirring, the temperature, the pH and the rate of uptake by
photosynthesis.

    The carbon in its various forms diffuses into the plant cells.
Diffusion occurs from a region of high concentration to one of lower
concentration and can be described by the equation:

$$F_j = - D_j C_j / x \qquad [2.1]$$

where F_j is the flux of substance j per unit area (moles $cm^{-2}s^{-1}$),
        C_j/x the change in concentration (moles cm^{-3}) of j in the
        distance x (cm) and D_j is the diffusion coefficient of J
        ($cm^2 \ s^{-1}$).

The movement of carbohydrates

The use of $-D_j$ is a convention to indicate that the movement is from a region of high to one of lower concentration.

An alternative method of describing the movement of carbon dioxide which has proved to be very useful is by making the flux per unit area (F_{CO_2} moles cm^{-2} s^{-1}) proportional to the difference in CO_2 concentration between the atmosphere and the sites of CO_2 fixation (moles cm^{-3}) and inversely proportional to a transport resistance (s cm^{-1}). Hence:

$$F_{CO_2} = C/R \qquad\qquad [2.2]$$

Alternatively, the resistance, R, is sometimes replaced by a conductivity K where $K = 1/R$ and hence:

$$F_{CO_2} = KC \qquad\qquad [2.3]$$

This description of diffusion ignores any interactions between diffusing species, say between carbon dioxide moving into and oxygen out of a leaf, and assumes movement through a uniform medium. Analysis has shown that, as a first approximation, interactions can be ignored, but any substance moving into or out of a plant cell has to pass through at least two phases in sequence; the external medium and the plant. Separate resistances can, therefore, be assigned to each stage of movement through the two phases and, since they are in series, the total resistance (R) is made up of the sum of the individual resistances. The resistance in the plant has a residual term which can be ascribed to intracellular transport and biochemical events which need not concern us.

In the smaller lower plants the sugar and other compounds which are produced by photosynthesis diffuse out of the chloroplasts and throughout the plant body. The situation is rather more complex in the larger plants which have evolved tissues specifically concerned with long distance transport. Yet further complications arise because of the different mechanisms of carbon fixation and storage. In this connection we can recognise three groups of plants; the C_3 plants, C_4 plants and CAM plants (those showing Crassulacean Acid Metabolism).

In C_3 plants the initial reduction of the CO_2 is catalysed by ribulose bisphosphate (RUBP) carboxylase in the chloroplasts of the mesophyll cells. Two molecules of phosphoglyceric acid are produced for each molecule of CO_2 reduced and are used to form hexoses and other substances which then enter into further reactions. Some of the hexoses are used to produce starch in the chloroplasts and this is laid down in the starch grains. Hexose molecules also diffuse out of the chloroplasts into the cytoplasm and produce sucrose and other materials. Some of the latter can enter into the photorespiratory cycles and can produce yet more sucrose.

In C_4 and CAM plants the initial site of CO_2 reduction is also the chloroplasts of the mesophyll cells but the enzyme involved is phosphoenolpyruvate (PEP) carboxylase. The products of fixation are malate in CAM plants and malate or aspartate in C_4 plants. In the C_4 plants the malate or aspartate then move into the chloroplasts of the bundle sheath cells where they are decarboxylated and the CO_2 produced is reduced via RUBP carboxylase and the pentose phosphate pathway. Starch and sucrose synthesis occur in the bundle sheath cells and if any CO_2 is produced by photorespiration it is reduced in the mesophyll before it can move out of the leaf. There is, therefore, in C_4 plants a spatial separation between the sites of initial CO_2 fixation and carbon storage. In CAM plants there is a time separation between these two processes. The CO_2 is reduced in the mesophyll cells in both the dark or light. The malate produced is decarboxylated in the light and the CO_2 refixed via the pentose phosphate pathway.

2.3 Movement of carbohydrates into the phloem

It will be obvious from the previous discussion that the pathways followed by assimilated carbon as it moves into the sieve tubes will vary with the photosynthetic pathway used by the plants. It should be noted, however, that the differences between C_3 and C_4 plants are not absolute. For example, in the C_4 plant *Digitaria sanguinalis* about 15 per cent of the carbon fixation may take place directly via the pentose pathway. Nevertheless, it is convenient to consider the movement of carbon into the phloem in these two groups of plants independently, even though some parts of the discussion are applicable to both groups.

2.3.1 Movement in C_3 plants
The exact pathway followed by sugar as it moves through the mesophyll into the sieve tubes is not clear. The most direct pathway would be through the plasmodesmata which pass from cell to cell, i.e. through the symplast. It has been suggested that this is the most important pathway in *Nicotiana tobaccum* and *Triticum aestivum*. In contrast, it is thought that in *Beta vulgaris* and *Vicia faba* the sugars move from the mesophyll cells into the apoplast and diffuse through this to the veins where they are absorbed by the phloem tissues (Fig. 2.1). There seems to be at least two possible modes of entry into the sieve tubes; either directly through the plasmalemma of the sieve tube – companion cell complex, or through transfer cells which in some

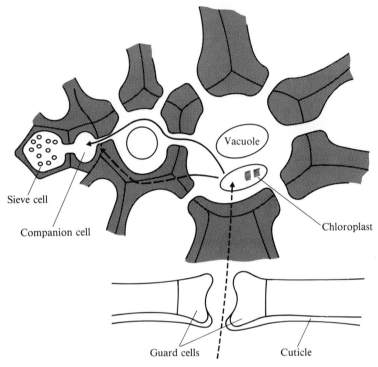

Fig. 2.1 A diagram showing the pathways followed by CO_2 (......) to the chloroplasts and sucrose from the chloroplasts to the companion cells and sieve tubes via the apoplast (- - - -) and symplast (——).

species are often associated with the phloem of minor leaf veins (see Fig. 1.23). The convoluted cell wall on the side away from the sieve tube, and its associated plasmalemma, presents a large surface area through which absorption can occur and the plasmodesmata facilitate transfer into the sieve tubes.

It would be naive to suppose that even within a single species transport into the sieve tubes occurs by only one of the pathways described. It is more likely that circumstances dictate which pathway predominates at any one time. Transport through the symplast seems quite feasible, with estimates of the fluxes of sucrose through a plasmodesma ranging from about 2.5×10^{-19} to 8×10^{-19} mol sucrose plasmodesma^{-1} s^{-1}. If this flux were achieved by the flow of a solution it would require the contents of a plasmodesma to be changed about seven times per second and this could be achieved by only a small difference in pressure between the two ends of the tube. If the movement was by diffusion the concentration difference required to achieve these rates would again be modest. The requirements for either mechanism would become more demanding, however, if the flux was greater or if movement was restricted to the desmotubule.

Transport of sugar through the alternative pathway, the apoplast, seems equally possible. Although no measurements are available, it might be expected that the concentration of sugars in the symplast would be higher than in the apoplast and that there would be a tendency for the sugars to diffuse into the latter. In such a situation the presence of mechanisms by which the leaf cells could accumulate sugars and other materials from the apoplast would be of distinct advantage to the plant and these have been demonstrated. Further, uptake experiments using radioactive tracers have shown that in petiolar tissue the rates of sucrose and phosphate uptake by vascular tissues are greater than by the surrounding parenchyma cells (Table 2.1). Autoradiography of the tissues used in these experiments

Table 2.1 Rates of uptake of sucrose and phosphate by fresh and aged tissues of celery petiole (μ mol g fresh wt^{-1} h^{-1}) (after R. Bieleski, 1966).

Tissue	Sucrose molarity				Phosphate molarity				
	10^{-2}	10^{-3}	10^{-4}	10^{-5}	10^{-2}	10^{-3}	10^{-4}	10^{-5}	10^{-6}
Fresh* celery petiole parenchyma	90	35	3.5	0.8	92	11	1.3	0.16	0.02
Aged* celery petiole parenchyma	346	120	19	2.9	225	147	51	16	2.0
Fresh* celery petiole vascular bundle	3080	670	74	10.4	1270	133	11.3	1.1	0.11
Aged* celery petiole vascular bundle	4480	1005	123	17	1700	560	210	60	7.1

* Tissues were kept in distilled water at 0 °C after collection and then rinsed. Fresh tissues were transferred to aerated 10^{-4}M CaSO$_4$ or CaCl$_2$ at 24°C for 10–20 min before use. This treatment was extended to 20–24 h for 'aged' tissues.

indicated that the most active sites of uptake in the vascular bundles were the sieve tubes. These observations, together with the ability of the phloem to accumulate materials against considerable concentration gradients by mechanisms which are sensitive to metabolic inhibitors, and which are promoted in the presence of ATP, all support the suggestion that there is an active 'loading' of material into the sieve tubes.

The diffusion of substances to the sieve tubes could be difficult in some circumstances because such movement would tend to be against the bulk flow of water in the opposite direction in the transpiration stream. This is, however, quite feasible since the two movements are brought about by different mechanisms. The transpirational movement is in response to differences in water potential between the xylem and the external atmosphere whereas the movement of sugar is probably by diffusion brought about by differences in sugar concentration. If the gradient in sucrose concentration is sufficiently great relative to the differences in water potential it would be quite possible to have the simultaneous movement of water and sugar in opposite directions through the same tissue. Some species have avoided the problem by a spatial separation of the two fluxes. For example, in *Triticum* 99 per cent of the water leaving the xylem is lost from the midrib and lateral bundles but these take up only 15 per cent of the sugar (Table 2.2). In contrast, little water is lost through the

Table 2.2 Comparison of the influx of sugar and efflux of water through the different types of vascular bundle in 1 cm² of wheat leaf (after J. Kuo, T. P. O'Brian and M. J. Canny, 1974).

Bundle type	Number in leaf	Sugar influx		Water efflux	
		p mol bundle^{-1} s^{-1}	p mol (cm² leaf)$^{-1}$ s^{-1}	p mol bundle^{-1} s^{-1}	p mol (cm² leaf)$^{-1}$ d^{-1}
Midrib	1	0.94	0.94	3.9×10^4	3.9×10^4
Large laterals	6	0.75	4.5	1.7×10^4	10.2×10^4
Small laterals	4	0.80	3.2	6.9×10^3	2.76×10^4
Large intermediates	13	2.01	26.1	83	0.1×10^4
Small intermediates	10	2.30	23.0	11	0.01×10^4

intermediate bundles which provide the major pathway of sugar uptake. There are two main reasons for this separation. The xylem cells in the small intermediate bundles have smaller diameters than those in the other bundles and hence offer a much larger resistance to the movement of water. Secondly, the phloem occupies a much greater proportion of the surface area of the intermediate bundles and therefore offers a larger area, and greater number of plasmodesmata, through which the sugar can move.

Irrespective of how the sugars are delivered to the sieve tubes they are always at a greater concentration in the sieve tubes than in the surrounding mesophyll cells. The result of this is that in *Beta vulgaris*, for example, the sieve tubes have a higher osmotic potential than the mesophyll cells; −2500 to −3000 J kg^{-1} and −1500 J kg^{-1} repectively. It is evident from this type of data that there must be some discontinuity in the transport pathway between the mesophyll cells and sieve tubes which allows these large concentration and osmotic differences to be maintained. If there were no symplastic connection between the mesophyll cells and phloem the differences could be ascribed to the operation of uptake mechanisms at the plasmalemmata of the sieve tubes and/or companion cells. In some species, however, plasmodesmata appear to connect the phloem and mesophyll and unless these can act as a type of valve it is difficult to see how the differences can be supported.

One proposed mechanism for the entry of sugars into the sieve cells is shown in Fig. 2.2. It is suggested that the observed high pH and

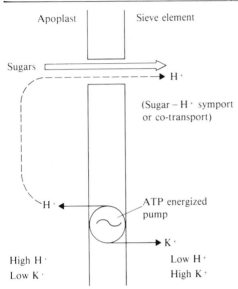

Fig. 2.2 A suggested scheme for an ATP energized pump causing the influx of potassium and the efflux of protons across the sieve cell plasma membrane and the co-transport of protons and sugars into the sieve cell down the resultant proton gradient (after F. Malek and D. A. Baker, 1977).

potassium ion concentration of sieve cell sap (Tables 2.5, 2.6 and 2.7) result from the operation of a pump at the plasmalemma of the sieve cell which produces a simultaneous efflux of hydrogen ions and influx of potassium ions. As a result of this a hydrogen ion gradient is maintained between the protoplast of the sieve cell and the surrounding apoplast and that there is a coupled transport of sugars and hydrogen ions down this gradient.

2.3.2 Movement in C₄ plants
The various chemical interconversions and inter- and intra-cellular movements associated with the production of starch and sucrose by C₄ plants have been discussed above. But there is much less information

about the movement of carbon into the sieve tubes of C_4 than C_3 plants. The plasmodesmata between the mesophyll and bundle sheath cells seem to be sufficiently numerous to accomodate the necessary flux of organic acids through the symplast. The decarboxylation of malate or aspartate in the bundle sheath cells can lead to CO_2 concentrations of about 25 μmol so ensuring a more than adequate supply of CO_2 for the RUBP carboxylase but what is not clear is whether a high concentration of CO_2 is retained within the bundle sheath cells, or whether refixation is immediate.

The rate of export of sugars from the leaves of C_4 plants is usually greater than that from leaves of C_3 plants. Hence, less radioactivity is retained in a leaf of the C_4 plant *Amaranthus* after a period of export than in the C_3 plant *Lycopersicon* (Fig. 2.3). This effect has been seen

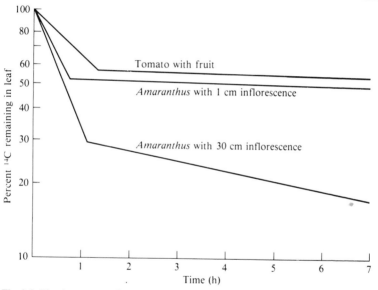

Fig. 2.3 The time course of the loss of ^{14}C-labelled assimilates from reproductive tomato and *Amaranthus* plants. The rate of export from the C_4 *Amaranthus* leaves is greater than from the C_3 tomato and is greater in the *Amaranthus* plants with large inflorescences (after J. Moorby and P. D. Jarman, 1975).

in several comparisons of C_3 and C_4 plants and has been ascribed to several factors. In *Amaranthus*, for example the atoms of ^{14}C have to pass, on average, through only 1.1 mesophyll cells before entering a sieve tube. In contrast, in *Lycopersicon*, the atoms have to move through about 5 mesophyll cells. There is, therefore, a greater amount of phloem per unit leaf area in the C_4 plant. Another suggestion, as yet unsubstantiated, is that compartmentation of the sugar in C_4 plants is such that it allows diffusion of the sugars down a concentration gradient rather than an active accumulation against a concentration gradient as in the C_3 plants.

2.3.3 The Kinetics of Movement into the Phloem

If a leaf is exposed to labelled CO_2 for a short time there is a period of a few minutes when the amount of label in the leaf shows little change. Diffusion of the CO_2 to the sites of initial fixation takes less than a second. Hence, the delay before the export of carbon starts must be ascribed to movement from the chloroplasts to the sieve cells. Thereafter, the amount of radioactivity in the area of leaf exposed to the labelled CO_2 starts to decline and typically follows a time course similar to that shown in Fig. 2.3 There is an initially rapid exponential loss of label from the leaf which is replaced after about 2 h by a second slower exponential loss. If the experiment is continued (Fig. 2.4) there appears to be yet a third slower exponential phase. In most experiments the respiratory component in the short-term loss of label from the leaf is small. Further, the experiment can be arranged so that

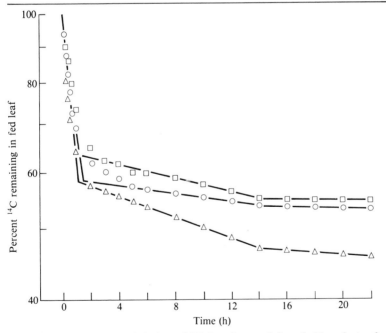

Fig. 2.4 The time course of the loss of ^{14}C from leaves of three fruiting plants of *Lycopersicon esculentum* after exposure of the plants to $^{14}CO_2$ for 5 min (after J. Moorby and P. D. Jarman, 1975).

41

the possibility of any loss by movement through the xylem is negligible and hence the reduction in counts can be ascribed to movement into and through the phloem. It was possible, therefore, to produce an estimate of the rate at which the labelled carbon entered the phloem of *Zea mays* by taking data similar to that shown in Figs. 2.3 and 2.4, differentiating with respect to time, and reversing the sign (Fig. 2.5).

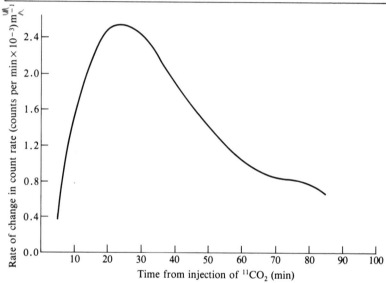

Fig. 2.5 The first derivative of a polynomial regression fitted to data similar to those in Figs. 2.3 and 2.4 but obtained using *Zea mays* which had been exposed to $^{11}CO_2$ for 1 min.
N.B. The sign is reversed to convert the rate of loss from the fed region to a rate of input into the phloem (after J. H. Troughton, J. Moorby and B. G. Currie, 1974).

The slow movement of carbon into the sieve cells led to the expansion of the 1 min pulse of labelled CO_2 used in this experiment to a Gaussian-shaped pulse of labelled assimilate entering the sieve tubes with a half width (width at half maximum height) of 30–40 min.

Similar delays could be seen when the distribution of ^{14}C was followed through the different tissues of *Vicia faba* leaves after exposure of the leaf to $^{14}CO_2$ for 4.5 min (Fig. 2.6). The amount of ^{14}C was initially high in the palisade mesophyll but rapidly declined. In contrast, the amounts in the spongy mesophyll and veins increased more slowly. The ^{14}C content of the spongy mesophyll declined more quickly than that of the veins and the general picture is of the gradual movement of the ^{14}C through the mesophyll and into the veins and the 4.5 min pulse of $^{14}CO_2$ was extended to a pulse of ^{14}C in the veins with a

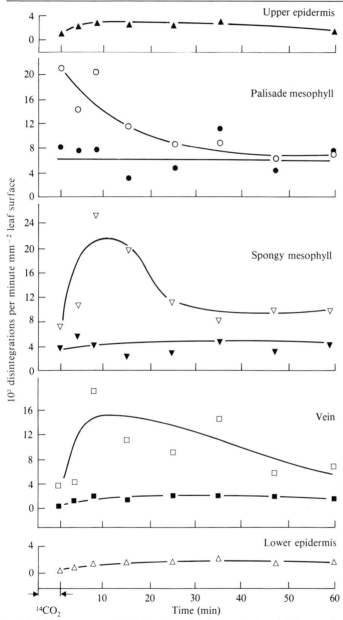

Fig. 2.6 The movement of ^{14}C through various tissues of *Vicia faba* leaves after exposure of the leaves to $^{14}CO_2$ for 4.5 min. Open and closed symbols show changes in the water soluble and insoluble fractions respectively (after W. H. Outlaw and D. B. Fisher, 1975).

43

half width of about 30 min. The insoluble fractions changed less rapidly than did the soluble and seemed to act as temporary stores of carbon. The delays in the movement of carbon into the sieve tubes cannot, therefore, be ascribed only to transport problems. The conversion of a proportion of the assimilated carbon to forms such as starch, proteins and cell material which are not immediately available to the translocation system will also impose delays on the export of the carbon.

The inter-relationships between these various forms and translocation can be described in terms of a three-compartment model (Fig. 2.7). In this, the carbon is presumed to enter and leave the leaf

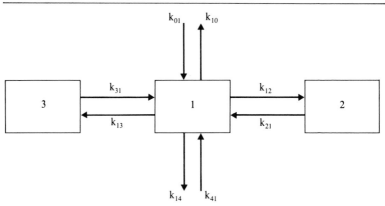

Fig. 2.7 A three compartment model which can be used to simulate the movement of carbon into and out of leaves. The flux from a compartment is given by the product of the size of the compartment (the number of carbon atoms) and the relevant exchange coefficient.

from a labile compartment. Exchange occurs between the labile compartment and two storage compartments (2) and (3); one has a residence time (the average period spent by a carbon atom in the compartment) of the order of two days whereas that of the second is several weeks. These compartments probably correspond with the short-term storage materials in the leaf such as starch, fructosans and proteins and the more permanent structural material. The rates of exchange between the compartments determine the pattern of export from the leaf. In experiments of less than 12 h duration it has proved to be unnecessary to consider the long-term storage compartment (3). The effect of light, darkness and CO_2 on the export of ^{14}C from leaves of *Amaranthus caudatus* and the presence and absence of fruit on export from leaves of *Lycopersicon esculentum* on the behaviour of the exchange coefficients of such a two-compartment model is shown in

Table 2.3 The effect of various treatments on the transfer coefficients of a 2 compartment model of a leaf (cf. Fig. 2.7) fitted to data similar to that shown in Figs. 2.3 and 2.4 (after J. Moorby and P. D. Jarman, 1975).

Species	Treatment	Transfer coefficients $(10^{-4}\ min^{-1})$		
		K_{13}	K_{12}	K_{21}
Amaranthus	Light, plus CO_2	140 ± 22	148 ± 21	4.1 ± 0.9
caudatus	Light, minus CO_2	51 ± 12	44 ± 9	7.0 ± 3
	Dark, plus CO_2	65 ± 11	55 ± 10	15.4 ± 3
	Dark, minus CO_2	59 ± 1	41 ± 12	11.4 ± 0.4
Lycopersicon	Plant with fruit			
esculentum	at 15–20 °C	69 ± 19	90 ± 34	4.7 ± 1.4
	Fruit removed	25.4 ± 5	51 ± 9	13.1 ± 4.6
	Plant with fruit			
	heated to 30 °C	84 ± 8	83 ± 25	3.2 ± 0.4

Table 2.3 The flux of carbon out of, or between, compartments is given by the product of the size of the compartment and the transfer coefficient. The data in Table 2.3 indicate, therefore, that when photosynthesis in *Amaranthus* is inhibited the rate of translocation from the leaf is reduced by about 60 per cent and the rate of mobilization of the stored carbon is increased by a factor of 2–3. In

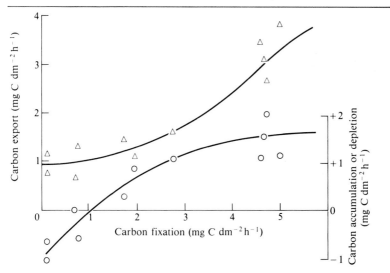

Fig. 2.8 The relationship between the rate of carbon fixation by leaves of *Lycopersicon esculentum* and the rates of carbon accumulation or depletion (○——○) and export (△——△) from the leaves (after L. C. Ho, 1976).

The movement of carbohydrates

Lycopersicon, fruit removal decreased the rate of translocation whereas heating the fruit increased the rate.

Thus, the ability of the plant to store carbon in the leaves which is surplus to the immediate requirements of translocation and leaf growth allows translocation to proceed when the current rate of photosynthesis is insufficient to meet the current demands. This was shown explicitly in other experiments on *Lycopersicon* in which the rate of translocation was maintained at 1 mg carbon dm^{-2} h^{-1} when the rate of photosynthesis was less than 1 mg carbon dm^{-2} h^{-1} (Fig. 2.8). The same buffering of the system can also be seen in Fig. 2.9 in the maintenance of translocation throughout the night and into the following photoperiod.

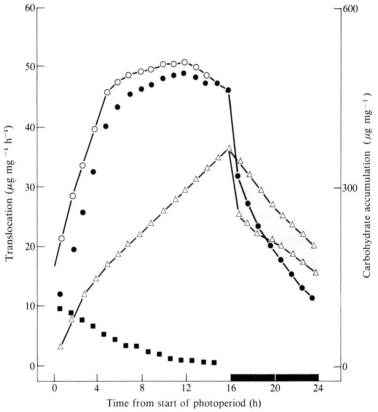

Fig. 2.9 The time course of translocation of carbohydrate produced during the current photoperiod (●) or the preceding photo-period (■) and the total amount translocated (○) and the carbohydrate accumulated in the leaf (△). The shaded area indicates the amount of carbohydrate lost by respiration during the dark period (after C. J. Pearson, 1974).

46

2.4 The kinetics of movement through the phloem

It is possible by a variety of techniques to show that as materials (ions and other compounds in addition to sugars), move through the phloem there is exchange between the sieve tubes and the surrounding tissues. Some of this lateral movement is obviously essential to supply nutrients to these tissues. There is also evidence of the temporary storage of materials outside the transport pathway which can be called on after only a short period, say during darkness, or after a longer time, for example the transfer of stem reserves into developing cereal grains or tubers of *Helianthus tuberosum* (see Chapter 4).

If a short pulse of labelled CO_2 is supplied to a leaf there is a delay before the label can be detected at positions along the stem of the plant. This delay is increased with distance from the leaf, and reflects the time required for the label to move from the leaf to the various positions (Fig. 2.10). As such, it is a function of both the distance travelled and the speed of movement and will be discussed in more detail below. The subsequent increase with time in the amount of label follows a sigmoid curve to a value which remains constant for a period (Fig. 2.10*a*) or increases, or decreases (Fig. 2.10*b*).

The simplest explanation of these results is that there is lateral movement of the label between the sieve tubes and the surrounding tissues. In situations where there is little movement back into the sieve tubes the count rate at a position will tend to remain constant whereas if a considerable amount of label is returned the count rate will decline. The data in Fig. 2.10*b* for example, show the change in the amount of labelled carbon at four positions on a leaf of *Triticum aestivum* after the assimilation of a pulse of labelled CO_2 in a distal part of the same leaf. The whole leaf was exporting carbon and hence the predominant direction of movement would have been into the sieve tubes and out of the leaf. In leaves where export is rapid the count rate can decline almost to zero.

Another aspect of the results in Fig. 2.10 is that the duration of the sigmoid parts of the curves tends to increase i.e. the pulse of label spreads as it moves through the plant. This can best be seen by differentiating the parts of the curves with positive slopes. The spreading of the pulse of label is shown by an increase in the half-width of plots of these differentials against time (Fig. 2.11). There are several possible reasons for this increase. For example, the lateral exchange of label between the sieve tubes and surrounding tissues would contribute to the spread as would the transport of label at a range of speeds. The first possibility has been discussed above and the second is highly likely.

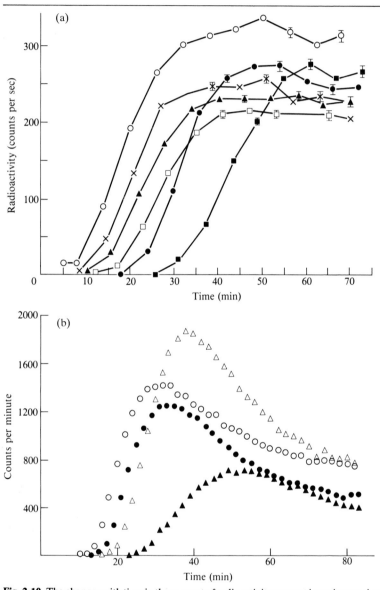

Fig. 2.10 The changes with time in the amount of radioactivity present in various regions distant from a leaf exposed to $^{11}CO_2$ for a short period.

(*a*) The ^{11}C present in sections of stem 10 cm (○——○), 14 cm (×——×), 18 cm (▲——▲), 22 cm (□——□), 26 cm (●——●) and 30 cm (■——■) from a leaf of *Glycine max* exposed to $^{11}CO_2$ for 3 min (after J. Moorby, M. Ebert and N. T. S. Evans, 1963).

Fig. 2.10 *(continued)*
(b) The ^{11}C present in sections of *Triticum aestivum* leaf 10 cm (○——○), 15 cm
(●——●), 20 cm (△——△) and 26 cm (▲——▲) from the area exposed to $^{11}CO_2$ for
3 min (unpublished data J. Moorby).

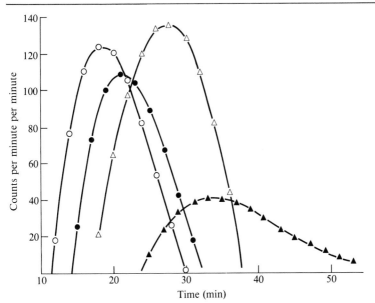

Fig. 2.11 The first differentials with respect to time of polynomial regressions fitted to
the parts of Fig. 2.10b with increasing count rates. These show how the half-widths of the
tracer fronts increased with distance travelled from the 3 min pulse of $^{11}CO_2$.

Except in very special circumstances any label which is introduced
into the phloem moves through the plant in more than one sieve tube,
and often in several vascular bundles. If transport through a sieve tube
is considered to be a flow process through a pipe of uniform
cross-section the volume flow rate Q (ml³ s⁻¹) is given by the Poiseuille
equation:

$$Q = \frac{(P_0 - P_1) \, \pi \, r^4}{8 \, \eta l} \qquad\qquad [2.4]$$

Where $(P_0 - P_1)$ is the difference in effective pressure over the length
of the pipe l (cm), r (cm) is the radius and η (g cm⁻¹ s⁻¹) the viscosity of
the fluid. The effective pressure difference includes any gravitational

component, i.e. $P = P + \rho gh$ where ρ is the density of the fluid, g (cm s^{-2}) is the gravitational constant and h (cm) is the elevation of the end of the pipe in question. Hence, the amount of solution moved will be very sensitive to the radius of the pipe and the speed of movement will be proportional to the fourth power of the radius. If the radius of the largest sieve tube in a bundle is only 50 per cent greater than that of the smallest, and the forces acting to produce the flow are identical, the speeds through the two would differ by a factor of 5. It is not surprising in these circumstances to find variations in the speeds at which materials move through plants. What is surprising is that the variation is not greater than is usually observed.

In theory, it is easy to estimate the speed of translocation. It is simply necessary to measure the time required for a tracer to move over a known distance. However, it will be obvious from the discussion above that it is difficult to characterize the tracer front. It is not sufficient to estimate the time of arrival of tracer at any one position. This will be dependent on the sensitivity of the detector and the geometry of the detection system at each position. Some attempt can be made to correct for the first, although it is impossible to correct for something which is undetected. Further, it is often impossible to produce really effective corrections for geometrical factors because of uncertainty about the site of the tracer in the tissue.

The method usually adopted has been to expose leaves of several plants to $^{14}CO_2$. These are harvested in succession, divided into sections and the amounts of ^{14}C determined at different distances along the transport pathway. The results are then expressed as the logarithms of the counts against distance travelled at a range of times (Fig. 2.12a). The speed of tracer movement can be estimated by plotting the time at which the count rate at any one position reaches a set value against the distance of the counting position from the source of label (Fig. 2.12b). The slope of this graph is an estimate of the speed of translocation, in the example shown it was 0.84 cm min^{-1}. Although, as can be seen, this method can provide very useful information, it ignores the effects of the lateral loss of tracer during translocation. These increasingly delay the time taken to attain any set count rate as the distance travelled increases. The speed will, therefore, be underestimated. Moreover, only one estimate of the

Fig. 2.12 (a) The changes with time in the count rates at different positions along the ▶ stem of *Glycine max* after the exposure of a leaf to $^{14}CO_2$. The positions, A, B, C and D indicate the distances along the transport path at which the count rate was 100 cpm at 10, 17, 25 and 35 min respectively (after D. B. Fisher, 1970).
(b) The displacement with time of a specific point on the ^{14}C front (100 cpm) as the front moved through a stem of *Glycine max*. The points are data obtained from Fig. 2.12a, the line is the regression $y = -2.47 + 0.84x$ ($r^2 = 0.96$) where y is distance moved (cm) and x time taken (min). The intercept on the abscissa was at 2.94 min.

speed is produced when, as was shown above, there are probably a large range of possible speeds.

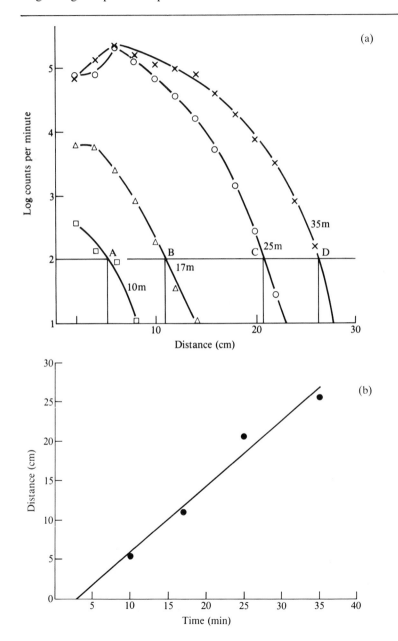

These problems can be reduced by the use of alternative methods. That described above is most suitable when ^{14}C is used as a tracer. It is possible to use ^{14}C to obtain data similar to that shown in Fig. 2.10, but it is very laborious because a different plant has to be used to obtain the data at each time interval because the ^{14}C emits weak particles which cannot easily be detected *in vivo*. Another isotope of carbon is ^{11}C. This emits positrons which in turn produce two annihilation quanta (γ-rays) which can be detected easily *in vivo* so permitting the simultaneous determination of the change in count rate at a number of positions. Further, the short half-title of ^{11}C (20.3 min) allows repeat experiments on the same plant. Unfortunately, the short half-life also restricts the use of ^{11}C because it has to be produced specially for each experiment and hence requires access to a suitable accelerator. All the results quoted in this section, except those in Fig. 2.12, have been obtained using ^{11}C. It can be seen that if the front of ^{11}C can be characterized in some way as it moves through the plant, its speed of movement can be calculated. Two methods have proved useful. Curves can be fitted to data similar to those in Fig. 2.10, and used to determine the times when 10, 20 . . . 90 per cent of the maximum count rate were attained (The 10 – 90 percentile points). The times taken for these points to move a known distance between the counting positions provides nine estimates of speed ranging over the whole tracer front (Table 2.4). An alternative estimate of the mean speed of movement can be obtained by using the time of maximum rate of change in count rate to indicate the centre of the front. This is done most easily by using the peak of the first differential of the fitted curves (Fig. 2.11). These values are also quoted in Table 2.4. It can be seen that in these experiments the ^{11}C at the beginning of the front appeared to be moving at almost twice the speed of that at the end of the front and the speed seemed to decrease with distance moved.

The speed of translocation in *Glycine* estimated by the methods shown in Figure 2.12 was much slower than that in *Zea* (Table 2.4). Although these estimates cannot be compared directly, they do illustrate what appears to be quite general phenomenon i.e. the mean speed of translocation is greater in C_4 than in C_3 plants. This, coupled with the greater proportion of assimilates which are exported from the leaves of C_4 plants (Fig. 2.3) suggests that the whole transport system may be more effective than in C_3 plants. However, the speed and amount moving are not the only criteria which have to be considered in assessing the effectiveness of transport systems to the growth of the plant. Equally important are the control of the direction of translocation and the utilization of the translocated material.

In several experiments attempts have been made to compare the relative speeds of translocation of sugars, water and various ions. None of these have been particularly successful because of the problems of

Table 2.4 Estimates of the speed of translocation at different positions within a ^{11}C front as it moved over two successive distances of 10 cm along a leaf of *Zea mays*. See text for description of methods used (after J. H. Troughton, J. Moorby and B. G. Currie, 1974).

Position within ^{11}C pulse. % of maximum count rate	Speed (cm min^{-1})	
	10–20 cm from source of ^{11}C	20–30 cm from source of ^{11}C
10	4.86	3.58
20	4.38	3.18
30	3.96	2.88
40	3.60	2.70
50	3.56	2.52
60	3.06	2.34
70	2.88	2.16
80	2.64	2.04
90	2.46	1.80
Mean	3.48	2.58
Peak of first derivative	3.66	2.70

estimating speeds discussed above. To these difficulties must be added others which arise when the movement of different substances is compared. For example, the rates of entry into the phloem are unlikely to be the same and this will complicate estimates of speed. Also, the only generally available radioactive isotope which can be used to label water is tritium. Unfortunately, this exchanges readily with hydrogen atoms on other molecules and it is difficult to be sure that any mobile tritium is being transported as water.

The simple concept of a speed of translocation carries with it the assumption that the contents of the sieve tubes are moving by a flow process, i.e. there is a bulk flow of solution. If this is so, the amount of material moving, the flux, Q (g cm^{-2} s^{-1}) is the product of the speed of movement of the solution, S (cm s^{-1}) and the concentration, C (g cm^{-3}):

$$Q = CS \qquad [2.5]$$

It is difficult to measure this concentration because of the problems of obtaining samples of the contents of small sieve tubes embedded in other tissues. A factor which has to some extent mitigated this problem is the positive pressure which exists in the sieve tubes. Because of this they tend to produce exudates when cut and these can be collected and analysed. The techniques used, and ease of exudate collection, vary with species, but it has proved possible to obtain samples from, for example, the trees *Quercus rubra* and *Fraxinus americana* by making cuts with a chisel and from the smaller *Ricinus communis* by using a razor blade. A variant on this technique is to collect the exudate from the inflorescence stalk of palms, e.g. *Yucca flaccida*, after removal of the inflorescence. In all these methods care must be taken to avoid contamination with exudation from cells other than the sieve tubes. The most precise method is to use an aphid as a means of sampling. These insects insert their stylet into a single sieve element. If then the insect is cut from the stylet the release of pressure allows the sap to exude and it can be collected in capillary tubing.

When collected with care the exudates appear to be good samples of the material moving through the phloem, even though they are produced as a result of puncturing what is essentially a closed system. All the methods produce a larger volume of exudates than can be accounted for by the emptying of the damaged sieve tubes. Moreover, under favourable conditions the rate of flow and the concentration of the exudate can often be maintained for long periods and ^{14}C-labelled assimilates can be detected in the exudate soon after exposure of a leaf to $^{14}CO_2$ (Fig. 2.13). The probable reason why flow eventually stops is the plugging of the sieve pores by callose but before this happens there are several effects which can distort the results. In *Ricinus*, for example, the incision into the sieve tubes appears to increase the speed of flow, possibly by providing a pathway with a smaller resistance to flow. The data in Fig. 2.13 show that the concentration of the *Ricinus* exudate can be maintained for several hours but in *Fraxinus* the concentrations of sugars can decrease by 20–40 per cent within 30 min of cutting, probably because of the osmotic movement of water into the sieve tubes from the surrounding tissues. There is also the possibility that puncturing the sieve tubes might enhance the rate of exchange of materials between the sieve tubes and other tissues.

Analyses of exudates from *Ricinus* and *Yucca* given in Tables 2.5, 2.6 and 2.7 show several similarities between the sieve tube sap from the two species. The dry matter content of both is high (10–20%); the major contributor to this, about 80 per cent of the total, being sucrose. The major cation present in potassium and the concentration of calcium is relatively low. The pH of both is about 8. This seems to be true of all phloem exudates and is a useful indicator of the absence of contamination with exudates from other tissues. Xylem exudates, for example, are usually about pH 6.

The major sugar is not always sucrose. In *Fraxinus* and *Cucurbita melopepo* there are also considerable amounts of raffinose and stachyose (sucrose plus one and two galactose units respectively). In other species sugar alcohols appear to be prominent, for example in *Malus sylvestris*, sorbitol, and the brown algae *Alaria marginata*,

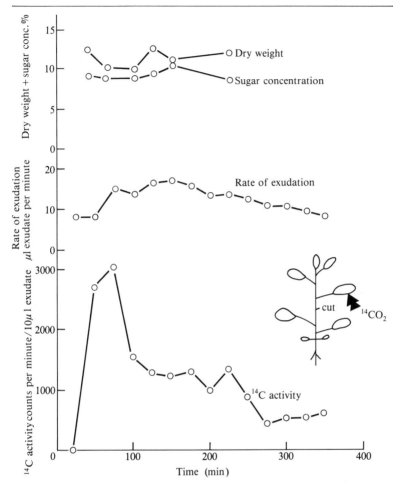

Fig. 2.13 The rate of exudation, concentrations of sugar and dry matter and ^{14}C content of the exudate from a 5 mm cut made into the phloem of *Ricinus communis* (after S. M. Hall, D. A. Baker and J. A. Milburn, 1971).

mannitol. Although monosaccharides have been detected, they are at a much lower concentration than the more complex sugars and the relative proportions suggest that they are breakdown products of the latter. The apparent discrimination against monosaccharides is thought to be a means of protecting the translocate from breakdown during transport. For example, invertase, which would aid the hydrolysis of sucrose, appears to be absent from the sieve tubes of *Robinia pseudoacacia*. Nevertheless, some breakdown does occur, and

The movement of carbohydrates

Table 2.5 The composition of exudate from the phloem of *Ricinus communis* expressed as mg ml^{-1} or m equiv. 1^{-1} or mM where appropriate (after S. M. Hall, and D. A. Baker, 1972).

Dry matter	100–125	
Sucrose	80–106	
Reducing sugars	Absent	
Protein	1.45–2.20	
Amino acids	5.2 (as glutamic acid)	35.2 mM
Keto acids	2.0–3.2 (as malic acid)	30–47 mequiv 1^{-1}
Phosphate	0.35–0.55	7.4–11.4 mequiv 1^{-1}
Sulphate	0.024–0.048	0.5–1.0 mequiv 1^{-1}
Chloride	0.355–0.675	10–19 mequiv 1^{-1}
Nitrate	Absent	
Bicarbonate	0.010	1.7 mequiv 1^{-1}
Potassium	2.3–4.4	60–112 mequiv 1^{-1}
Sodium	0.046–0.276	2–12 mequiv 1^{-1}
Calcium	0.020–0.093	1.0–4.6 mequiv 1^{-1}
Magnesium	0.109–0.122	9–10 mequiv 1^{-1}
Ammonium	0.029	1.6 mequiv 1^{-1}
Auxin	10.5×10^{-6}	0.60×10^{-4}mM
Gibberellin	2.3×10^{-6}	0.67×10^{-5}mM
Cytokinin	10.8×10^{-6}	0.52–10^{-4}mM
ATP	0.24–0.36	0.40–0.60 mM

pH	8.0–8.2
Conductance	13.2 micromhos cm^{-1} at 18 °C
Solute potential	−14.2 to −15.2 bars
Viscosity	1.34 cP at 20 °C

Table 2.6 The amino acid composition of a sample of exudate from the phloem of *Ricinus communis*. (Original data from S. M. Hall and D. A. Baker, 1972, recalculated to facilitate comparison with data in Table 2.5.)

Amino acid	Concentration mM	% of total amount of amino acids
Glutamic acid	13.00	34.76
Aspartic acid	8.80	23.53
Threonine	5.40	14.44
Glycine	2.40·	6.42
Alanine	2.00	5.35
Serine	1.60	4.28
Valine	1.60	4.28
Isoleucine	1.00	2.67
Phenylalanine	0.60	1.60
Histidine	0.40	1.07
Leucine	0.40	1.07
Lysine	0.30	0.53
Arginine	trace	–
Methionine	trace	–
Total	37.4	100

Table 2.7 The composition of exudate from inflorescence stalks of *Yucca flaccida* (after P. M. L. Tammes, and J. Van Die, 1964).

The substances listed in the first part of the table were determined in portions of freshly tapped exudate obtained a few hours after the removal of a slice of tissue from the cut inflorescent stalk.

The substances in the second part of the table were determined in a large volume (ca. 100 ml) of exudate collected over a period of 10 days.

total dry matter	17.1 – 19.2%
electric conductivity (20 °C)	1.03 mMho cm^{-1}
pH	8.0 – 8.2
sucrose	150 – 165 mg ml^{-1}
fructose	2 – 4 mg ml^{-1}
glucose	2 – 4 mg ml^{-1}
glucose-l-phosphate	*ca.* 1 mg ml^{-1}
total amino acids (as glutamine)	6.3 – 10.1 mg ml^{-1}
total protein	0.5 – 0.8 mg ml^{-1}
invertase	absent
total phosphorus	0.310 mg ml^{-1}
inorganic phosphorus	0.105 mg ml^{-1}
nitrate	absent
magnesium	0.051 mg ml^{-1}
calcium	0.014 mg ml^{-1}
potassium	1.680 mg ml^{-1}
sodium	0.004 1 mg ml^{-1}
zinc	0.002 1 mg ml^{-1}
iron	0.001 4 mg ml^{-1}
manganese	0.000 5 mg ml^{-1}
copper	0.000 4 mg ml^{-1}
molybdenum	0.000 01 mg ml^{-1}

the enzymes necessary for glycolysis appear to be present in the sieve tubes together with considerable amounts of ATP. The use of tracer techniques has confirmed this breakdown of mobile sugars although the actual cells in which it occurs are unknown. Some workers have suggested the companion cells as the most likely sites.

The concentrations quoted in Tables 2.5, 2.6 and 2.7 are mean values. They vary with the time of day and position within the plants. The concentration of sugar in the phloem increases during the day as photosynthesis produces carbohydrates and is probably emphasized by water loss as transpiration reduces the water potential of the xylem. Hence, the increase in sugar concentration in the phloem follows that in the leaves and is not seen in the woody tissues (Fig. 2.14). Data of this type was one of the first demonstrations that the phloem was the major translocation pathway for sugars. The diurnal changes can be seen most clearly when translocation takes place from a restricted source over a long distance; for example in a tree with a distinct crown (Fig. 2.15). Here the wave of increased concentration of sugar could be followed as it passed down the stem.

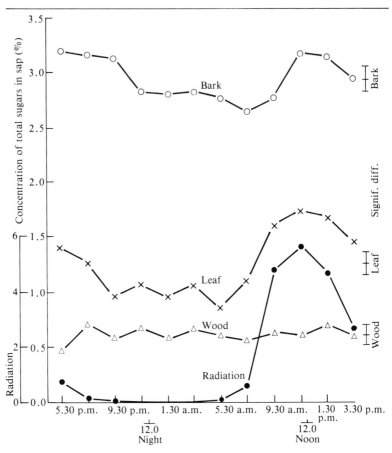

Fig. 2.14 The concentration of sugars in the sap of leaf tissue, bark, i.e. phloem and wood of *Gossypium barbadense* plants. The radiation values are in arbitrary units. The values for the significant differences of the analytical values are shown on the right (after T. G. Mason and E. J. Maskell, 1928*a*).

Several techniques have been used to estimate the rate of translocation, i.e. the amount of material moved per unit area of phloem per unit time. One method which has been used often has been to measure the dry weight increase over a known period of an organ such as a potato tuber or a cucurbit fruit. Since all the carbon stored by organs such as these has to be translocated through the phloem of the stolon or peduncle it is possible to calculate an average rate of translocation over the period of measurement. Work of this type has usually produced estimates in the range 1 to 2 g h^{-1} cm^{-2} phloem with

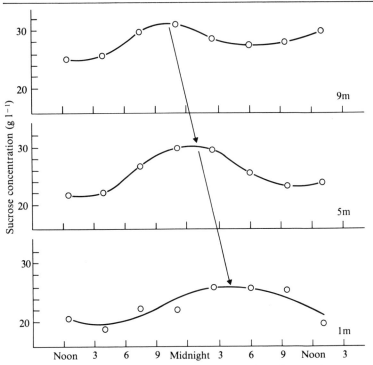

Fig. 2.15 The mean diurnal fluctuations in the concentrations of sucrose from cuts made at three heights, 1, 5 and 9 ms into the bark of two trees of *Fraxinus americana* (after M. H. Zimmerman, 1958).

some values of up to $4 \, g \, h^{-1} \, cm^{-2}$ phloem. It should be noted, however, that many of the early estimates did not allow for any respiratory losses and could, therefore, have underestimated the true value by up to about 30 per cent. Recent experiments, in which wheat plants have been induced to develop a complete root system from a single seminal root, suggest that the rate of assimilate translocation through the base of this root might attain values up to 10 times those quoted above. If these estimates are confirmed, they will raise many problems regarding the feasibility of some of the hypotheses of the mechanism of phloem transport (see section 2.5).

An alternative method of estimating the rate of translocation is to couple measurements of the rate of CO_2 exchange with changes in the carbon content of the leaf. The difference between the net amount of carbon assimilated by the leaf and that remaining in the leaf after a period of translocation provides an estimate of the rate of export of carbon from the leaf. This technique was used to obtain the data

59

shown in Fig. 2.8. The maximum rate in this data is equivalent to approximately 4.75 g dry wt h^{-1} cm^{-2} phloem.

A variation on this last method is to pass over a leaf a stream of air containing a known, fixed, concentration of CO_2 labelled with $^{14}CO_2$ at a constant specific activity. After a period to allow the carbohydrates in the source leaf to attain a constant specific activity (this is usually less than that of the $^{14}CO_2$ because of turnover of the polysaccharides produced prior to the start of the experiment) the rate of transport of assimilates into a sink, e.g. a young leaf, can then be estimated from the rate of arrival of ^{14}C in the sink and the specific activity of the source leaf material. One such series of experiments using sugar beet produced an estimate of 1.4 g h^{-1} cm^{-2} phloem.

It can be seen from the foregoing discussion that, with one exception, there is relatively little variation in the different estimates of the rate of translocation; most estimates varying by less than a factor of about four. Other experiments have also shown variations in the speed of translocation (Table 2.4). It would be useful, therefore, in view of the relationship between these two aspects of translocation described by equation [2.5] if it was possible to determine simultaneously the rate and speed of translocation and the proportionality factor relating them, the concentration. Unfortunately, this has not been possible, but simultaneous estimates of speed and rate usually produce calculated values of concentrations in the range 5 to 20 per cent sucrose; i.e. similar to independent estimates of sugar concentrations. Also consistent are simultaneous determinations of rate and concentration, which produce estimates of speed of between 40 and 100 cm h^{-1}. These indirect estimates are, therefore, consistent with direct observations of the three factors made at different times, or on different tissues, and thus give some confidence in the values. They do not, however, necessarily confirm the assumption implicit in equation [2.5] that there is a bulk flow of solution. It seems relevant, therefore, at this point to discuss whether there is a flow of solution through the sieve cells or whether movement is by some other mechanism.

2.5 The possible, and sometimes impossible, mechanisms of phloem movement

Thus far the movement into and through the phloem has been described, but there has been no discussion of the mechanism or mechanisms by which the material moves through the sieve cells.

Any explanation of the mechanism of sucrose transport through the phloem has to accommodate these generally agreed facts:

1. Longitudinal movement takes place through the sieve cells.
2. The speed of movement can vary within the range 30 to 400 cm h^{-1}.
3. The rate of movement can vary between 1 and 4 g cm^{-2} phloem h^{-1}.

These last two are quite large limits, but they are, nevertheless, sufficient to exclude some of the proposed mechanisms. In addition, the suggestions have to be compatible with the ways in which ions and water are moved.

It must be stated at the outset that until there is a generally agreed interpretation of the structure of sieve cells it will be impossible for there to be any agreement concerning the mechanism, or mechanisms, of transport through these cells. In this discussion, therefore, the pros and cons of the various theories will be briefly listed, together with considerations of how they can be reconciled with the different interpretations of phloem structure. It is not intended that the discussion be comprehensive. Although the elucidation of the mechanisms involved has formed a constant and intriguing theme in the study of phloem it has perhaps obscured the fact that the most important aspects of phloem physiology in terms of the control of plant growth are how much material can be transported from sources to sinks and how this amount and the direction of translocation are controlled.

One of the first widely accepted hypotheses concerning the mechanism of movement through the sieve tubes was developed by Münch. He suggested that there was an osmotically generated flow of solution. In its simplest form this hypothesis assumes that there is a high concentration of sucrose in the phloem of the source and that the plasmamembrane of the sieve cells is semipermeable. Water therefore moves osmotically into the phloem from the surrounding tissues. Because the cell walls prevent unlimited cell expansion the pressure in the sieve cells is increased and the solution in the sieve cells moves out of the source through these cells. In the sink the events are reversed. Sucrose moves out of the sieve cells into the surrounding tissues as does the water. The water is then either used in the sink or recycled in the xylem to other parts of the plant. The basic suggestion is, therefore, that there is a flow of solution through the sieve cells in response to an osmotically generated difference in pressure between the source and sink.

The concentration gradient between source and sink necessary for the operation of the pressure flow theory has been demonstrated on several occasions (see section 2.5.1). However, in a long series of experiments, Mason and Maskell went further and showed that the rate of translocation of sugars through cotton plants was proportional to the concentration gradient (Fig. 2.16) and that the situation could be described by the standard diffusion equation (equation [2.1]). When the appropriate values were inserted in the equation a value was

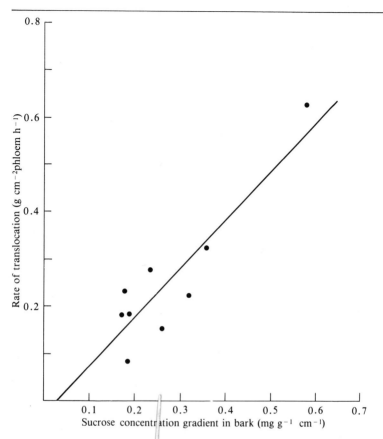

Fig. 2.16 The relationship between the rate of translocation in cotton plants and the gradient of sucrose concentration in the bark (after T. G. Mason and E. J. Maskell, 1928*b*).

obtained for *D* which was approximately 2×10^4 to 4×10^4 times greater than the diffusion coefficient for sucrose in water. It is, therefore, highly improbable that the movement was by straightforward diffusion. No one has attempted to repeat the experiments of Mason and Maskell, but, equally, no one has questioned their results. Indeed, a recent calculation from much less comprehensive data supports their value for *D*. Because of the improbability of translocation being by simple diffusion it has been suggested that there is an 'accelerated' diffusion; the acceleration usually being brought about by metabolic powered pumps. The pros and cons of several of these, and the pressure flow will now be discussed.

2.5.1 The pressure flow theory

It is clear that the simple form of the theory outlined above is too superficial. Recent accounts accept that there may be expenditure of metabolic energy in the loading of materials into sieve cells. Further, the lateral exchange of solutes between the sieve cells and surrounding tissues must mean that the plasmamembrane is not truly semipermeable, but merely more permeable to water than the solutes.

It is still a central tenet of the theory that the movement through the sieve cells is passive and takes place because of the osmotically generated pressure gradient. This gradient dP/dx can be described by:

$$dP/dx = RT\Sigma dC_i/dx + d\Psi/dx - 1/L_p.dJ_w/dx \qquad [2.6]$$

where R is the gas constant, T the absolute temperature and dC_i/dx is the gradient in solute concentration along the pathway in the sieve tubes. The last two terms in equation [2.6] describe the effects of the water potentials in the apoplasm which will affect the flux of water through the sieve tubes. Here, $d\Psi/dx$ is the gradient in water potential outside the pathway, dJ_w/dx the gradient in water volume flux into the sieve cells along the pathway and L_p is the hydraulic conductivity of the plasmamembrane of the sieve cells. If L_p is large, or if there is little change in J_w along the pathway, or both, the last term in equation [2.6] can be ignored.

The influence of the various factors indicated in equation [2.6] on the translocation of sugars seems to be satisfied at least qualitatively. Hence, it is possible to detect gradients in sugar concentrations along the translocation pathway (see Fig. 2.15 and, more clearly, Fig. 2.17). Experiments in which the water potential of the xylem was reduced by introducing mannitol solutions led to a reduction in the rate of exudation from aphid stylets inserted into adjacent bark and a concomitant rise in the concentration of sucrose in the exudate (Fig. 2.18). There have been no direct measurements of, say, the effects of changes in concentration gradient on the turgor pressure gradient in the sieve cells. The few attempts to measure dP/dx directly have shown that gradients exist and that pressures are greater in the sources than the sinks. For example, they are greater in higher parts of trees than in lower and in the source regions of herbaceous plants than in sinks initiated by either darkening or defoliating parts of the plants.

Although the observed pressure gradients are in the expected direction, the successful operation of a pressure flow mechanism is dependent on the presence of gradients sufficient to sustain the flow of solution at the rates and speeds observed. It is possible to derive relationships between the applied pressure and rate of flow of solution through a tube (equation [2.4]) and the speed of flow. The expression for the latter is:

$$dP/dx = 8\eta v/r^2 \times 10^5 \qquad [2.7]$$

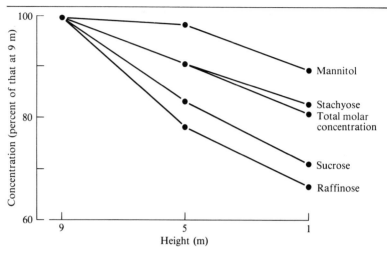

Fig. 2.17 Gradients in the concentrations of various constituents of sieve tube exudate from *Fraxinus americana* expressed as percentages of the concentrations at 9 m which were: sucrose, 82 mM; raffinose, 73 mM; stachyose, 202 mM, mannitol, 178 mM (after M. H. Zimmerman, 1957).

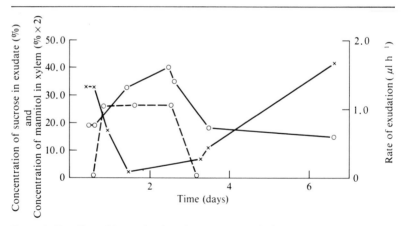

Fig. 2.18 The effect of decreasing the xylem water potential for a period, by passing a 15 per cent solution of mannitol through the xylem of *Salix viminalis*, on the rate of exudation from aphid stylets (×——×) and the concentration of sucrose in the exudate (○——○). The concentration of mannitol in the xylem exudate is also shown (○----○) (after P. E. Weatherley, A. J. Peel and G. P. Hill, 1959).

The various abbreviations in the equation are those used earlier. The major problem is to know what value to assign to r. If the radius of the sieve cell lumen is used, say 12 μm, together with a viscosity equivalent to that of a 10 per cent solution of sucrose, the pressure gradient required to generate a speed of movement of 1–2 cm min^{-1} is of the same order as would be generated by the observed concentration gradients. Unfortunately, however, there have been insufficient measurements of turgor pressure gradients to know whether the movement can be described by equation [2.7]. Other considerations would suggest that it is unrealistic to use the average radius of the sieve cell lumen because the flow will be dependent on the size of the sieve pores. These are typically about 5 μm long with a radius of 2.5 μm and a total area equal to about half the total area of the sieve plate. The speed of movement through the pore will, therefore, be twice that through the sieve cell lumen and the pressure gradient required to produce the same speed will be more than doubled. It is greater than might be expected from the reduced area because of relatively larger edge effects. Using these assumptions, it can be calculated that the concentration gradients found in trees could support transport through sieve cells with open pores over a distance of about 25 m; transport over longer distances would be unlikely.

These calculations are based on the assumption that the sieve cells and pores of the sieve plates are unobstructed. If there are any obstructions, the pressure required to generate a mass flow is increased and soon becomes excessive. The widespread occurrence of p-protein or other filamentous structures in sieve cells is, therefore, a major obstacle to the general acceptance of the mass flow theory and there appears to be a paradox. The speed of transport observed is orders of magnitude greater than can be accounted for by diffusion and yet the structure of the sieve cells appears to preclude the operation of a simple pressure generated bulk flow of solution. It is because of this apparent paradox that the reconciliation of the structural observations with the hydraulic requirements of the system has become the nub of the problem of the mechanism of movement. This problem is unlikely to be solved until techniques are developed which can show unequivocally whether the various structures and inclusions found in functioning sieve cells are real or artefacts.

There could be yet another problem associated with the acceptance of the pressure flow theory. If movement through the sieve cells is in response to a pressure gradient the flow should be in one direction only. There have been many experiments which have attempted to test this proposition. It is relatively easy to demonstrate the transport of, say, ^{14}C-labelled assimilates and ^{32}P in opposite directions through an internode or petiole. This is an insufficiently critical test, however, because transport of the two substances could be in separate vascular bundles. Even if transport was restricted to a single vascular bundle, it can be postulated that different sieve cells in one bundle could transport material in opposite directions. This explanation is less acceptable, however, because it is difficult to see how turgor pressure gradients could be maintained in opposite directions in adjacent files of sieve cells in the same vascular bundle. Nevertheless, the crucial test would be to demonstrate movement in opposite directions in a single sieve cell; two experiments appear to satisfy this requirement (to some people). It has proved possible to obtain exudate samples from an aphid stylet which contained two isotopes which had been applied to the plant on opposite sides of the stylet. The exudate undoubtedly came from one sieve cell, but the possibility exists that the isotopes entered the cell as a result of the release of pressure as the stylet penetrated the sieve cell. More convincing evidence for bi-directional transport comes from autoradiographic studies of transport through the petioles of *Cucurbita melopepo*. These showed the presence in single sieve cells of both ^{3}H-labelled and ^{14}C-labelled sugars which had been applied to two different leaves. The techniques used in these experiments were such that it is difficult to avoid the conclusion that there was bi-directional transport through one sieve cell.

We are left with the conclusion that although many observations support the idea that translocation through the sieve cells is by means of a pressure flow mechanism, others raise sufficient doubt to prevent the completely unreserved acceptance of the proposal. It is because of these doubts that various alternatives to the pressure flow hypothesis have been made. Several of these will now be discussed.

2.5.2 Active mechanisms

All of the active mechanisms which have been proposed attempt to explain several problems, some of which are difficult, if not impossible, to explain on the pressure flow hypothesis. These are the function, if any, of the structures which have been found in the sieve cells of most species and the relationship between the concentration gradient between source and sink and the rate of translocation.

Many of the proposed mechanisms attempt to assign a purpose to the p-protein and other structures by suggesting that they can act as pumps powered by metabolic energy. A prime requirement of any of these proposals is, therefore, that the metabolic processes in the sieve cells can supply sufficient energy to sustain the necessary rates of movement. Further, since no process is 100 per cent efficient, the metabolism should be capable of supplying more than the minimum amount of energy theoretically required.

The description of sieve cells provided in Chapter 1 suggest that they are metabolically active and this is supported by analyses which show the presence of ATP and a wide spectrum of enzymes. Moreover, they are in close association with cells which have very dense cytoplasm containing many mitochondria and other subcellular particles, e.g. the companion cells in angiosperms and the albuminous cells in the gymnosperms. It is impossible to measure the rate of respiration of sieve cells *per se*, but isolated phloem tissue has quite high rates of respiration. A consumption of 230 μl O_2 (g fresh wt of phloem)$^{-1}$ h^{-1} being an average value, equivalent to an energy supply of about 4.86 J (g fresh wt of phloem)$^{-1}$ h^{-1}. Hence, allowing for 30 per cent of this energy to be used for maintenance of the phloem structure, and a 50 per cent efficiency of conversion of the remaining energy, there remains in excess of 1.5 J (g fresh wt of phloem)$^{-1}$ h^{-1} available to provide the motive power for the translocation process. Calculations suggest that this could produce the same flow as pressure gradients of the order of 2.5×10^6 Pa m^{-1}, i.e. adequate to account for transport through most plants.

Although these calculations suggest that sufficient metabolic energy is available, they say nothing about how the energy might be converted into translocation. If there is a large active component in translocation it might be expected that cooling part of the transport pathway would inhibit translocation. This type of experiment has been done many times, usually with the same result. The initial response is an inhibition of translocation, but this is not permanent and after less than an hour translocation is resumed (Fig. 2.19). When maintained for long periods the low temperature appears to have no effect (Fig. 4.8).

The low temperature treatment might be expected to have reduced the rate of respiration and hence the energy available to power translocation. Another effect of the decreased temperature is to increase the viscosity of the contents of the sieve tube and this, again, would tend to inhibit translocation (equation [2.4]). However, if the low temperature treatment is restricted to the transport pathway and the source and sink regions are unaffected these might be expected to continue to function normally. The result, therefore, would be to increase the concentration gradient between the two ends of the treated region; an effect which might eventually compensate for the increased viscosity. Results of this type can, therefore, be interpreted as supporting the idea of a pressure generated flow of solution through the sieve cells. However, completely unequivocal evidence in support of this conclusion is still not available and hence it is still useful to consider the ways in which an active mechanism might operate.

Most of the proposals make a virtue of necessity by assigning some functional role to the structures seen in the sieve cells; the nature of the role usually being determined by the proposers interpretation of the structures seen; or vice versa. For example, one suggestion has been that membrane-bound transcellular strands exist in sieve cells, that these strands pass from cell to cell through the sieve pores, and that transport in opposite directions can take place in separate strands

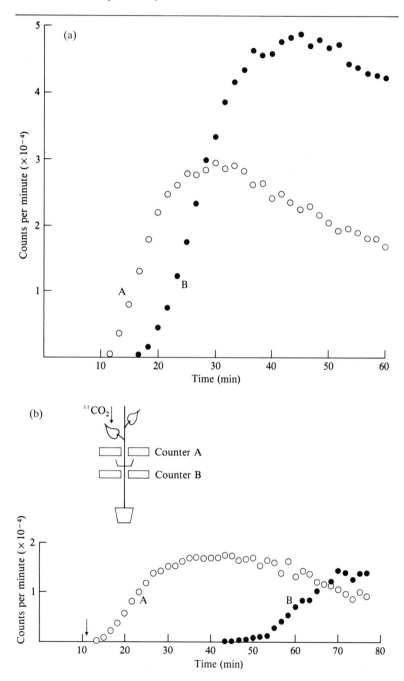

◀ **Fig. 2.19** The effect of cooling the stem of a sunflower plant on the translocation of
^{11}C-labelled assimilates. (*a*) Control data, counters A and B 11 and 23 cm from the fed
leaf respectively.
(*b*) Ice and water were put in the cup between the counters 11 min after exposure to
$^{11}CO_2$ (arrow). N.B. The lower maximum count rate in (*b*) may have been caused by
exposure to less $^{11}CO_2$ than in (*a*), but this would not have affected the time of arrival of
the ^{11}C at counter B.

through a single sieve cell (Fig. 2.20). It is proposed that flow through
the strands is initiated by means of a peristaltic-type mechanism.
Energetically, such a mechanism is feasible, and it could produce
tracer profiles akin to those produced by simple diffusion. Unfortunately, although lineally orientated structures seem to be present

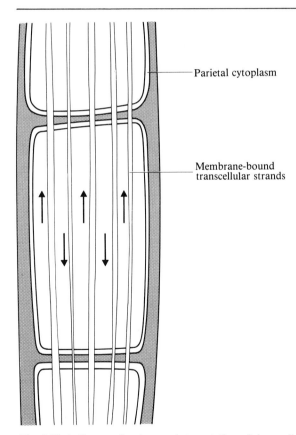

Parietal cytoplasm

Membrane-bound
transcellular strands

Fig. 2.20 A diagram showing one interpretation of sieve cell structure with
membrane-bound transcellular strands through which transport can take place in
opposite directions in the same cell.

in sieve cells there is no generally accepted unequivocal evidence that they are surrounded by membranes.

It is for the last reason that other workers suggest that the strands are made up of an association of smaller fibrils to form some sort of contractile structure. The nation of the association proposed varies from author to author and reflects the lack of unequivocal evidence (Fig. 2.21). The contractile structure could be the basic transport system with the mobile material being passed along the fibrillar material. In one regard this proposal is akin to an early suggestion that translocation takes place as an interfacial movement along surfaces in the cell. Because such a movement usually takes place in a monomolecular layer, objectors to the original idea have stated that it would result in an unacceptably small rate of translocation. This need not necessarily follow because the presence of strands etc. would significantly increase the area available for transport. In addition, the operation of any contractile forces would also increase the potential rate. A variation on this proposal is that the contractions in the strands could initiate a bulk flow in the solution surrounding the strands, i.e. there would be a pumping system extending throughout the sieve cells.

A system capable of producing movements of this type might be expected to be similar to the actin-myosin type of association found in muscle fibres and the cytoplasmic streaming systems of the slime moulds and amoebae. Such a system would be feasible on energetic grounds, but only few attempts to identify actin in p-protein have been successful and there is, as yet, no general agreement on how it could operate.

It will be seen from this discussion that we must revert to the conclusion reached at the end of the previous section. The active mechanisms proposed, like pressure flow, can account for some, but not all, the observed facts. Further, much of the debate has been conducted on the basis that active mechanisms and pressure flow are mutually exclusive. This need not be so, and there have been some recent suggestions that a pressure-type flow can be initiated by active processes. This possibility has obvious attractions but acceptance of the proposals will have to await further work.

Fig. 2.21 Further interpretations of sieve cell structure. (*a*) and (*b*) show ways in which ▶ the filamentous material could be attached to either cell walls or axial strands. It has been postulated that vibrations of the filamentous material might initiate a mass flow of solution.
(*c*) Yet another proposal where the axial filaments are capable of a peristaltic pumping (shown in more detail and arrowed in (*d*)). In addition, the filamentous material might initiate a flow of solution (after, Fensom, 1975).

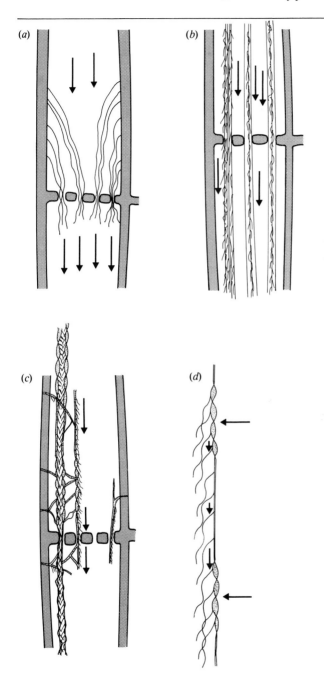

2.6 Movement out of phloem

It has been shown that in a leaf the concentration of sugar in the sieve cells is greater than that of the surrounding cells. The same appears to be true in other plant organs and, therefore, there will tend to be a diffusion of sugar out of the sieve cells which is countered by the loading process. This exchange, which occurs along most of the translocation pathway has been discussed above. Indeed, the continued existence of the plant necessitates that outside the source regions there is a net loss of nutrients from the sieve cells.

Those tissues in which there is a very large loss constitute the 'sinks' of the plant. It must be emphasized, however, that the division of the translocation system into sources and sinks is arbitrary. A tissue is only a source or sink in the sense that there is a large *net* export or import of material during the period in question. The situation can, and often does, change with time. These changes, and the interrelationships between sources and sinks, are discussed further in Chapter 4.

The initial movement of materials out of the sieve cells will be into the surrounding apoplast. The continued movement is dependent on the maintenance of a concentration gradient into the cells around the sieve cells; three factors can contribute to this. The growth and respiration of the surrounding tissues are of obvious importance since they use some of the carbohydrates and ions supplied. The second process, especially important in storage organs, is the production of reserve material such as starch or fructosans, proteins and fats. These substances are not transported through the phloem but their synthesis, will reduce the concentration of the simpler materials which are transported through and out of the sieve cells. The final method is important in those tissues which store sucrose at concentrations greater than those usually found in the sieve tubes. For example, the concentration of sucrose in both sugar beet roots and sugar cane stems can reach 12–16 per cent of the fresh weight or 70 per cent of the dry weight compared with the concentration in exudates from sieve cells of about 10 per cent. It can be seen that the movements out of the sieve cells can only be maintained if the sucrose is stored in a compartment isolated in some way from the free space. Further, it seems probable that there will be an active step at some point in the transport into the storage compartment. The situation has been studied in detail in sugar cane and the processes which appear to be involved are shown in Fig. 2.22. In this scheme the sucrose which diffuses out of the sieve cells is hydrolysed to glucose and fructose by an acid invertase in the apoplast. These monosaccharides are transported into a metabolic compartment in the parenchyma cells by a carrier-mediated active mechanism and are phosphorylated. Sucrose phosphate is then produced from the phosphorylated monosaccharides and the sucrose

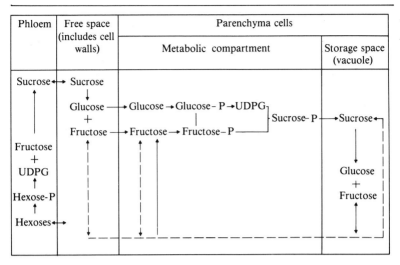

Fig. 2.22 A scheme suggesting the way in which sugar moves from sieve cells to the storage parenchyma in the stem of sugar cane. For explanation see text (after K. T. Glasziou and K. R. Gaylor 1972).

moiety of the sucrose phosphate is transferred to the storage compartment, probably the vacuole. Because of the high concentration of sucrose in the storage compartment there will necessarily be a back-diffusion into the metabolic compartment. It is hydrolysed there by a neutral invertase and the monosaccharides rephosphorylated.

The relative importance of these three mechanisms will vary from tissue to tissue and at different times in the same tissue. In many fruits, for example those of *Zea mays* and *Pisum sativum*, there is a period of rapid growth followed by one of sucrose accumulation and finally starch synthesis. The situation can become even more complicated in those tissues in which the activity of some of the enzymes involved in the production of storage compounds appear to be dependent on the continued supply of materials through the phloem. Here, interruptions in the supply of materials can lead to abnormal growth, e.g. chain tuberization in *Solanum tuberosum*.

2.7 Mathematical models of translocation

Various mathematical models of translocation have been produced based on both the pressure flow and other theories. They all basically assume the movement of sucrose or a sucrose solution through a permeable pipe, or series of pipes, with exchange of solutes between the mobile and stationary phases. Depending on the model in question, the compartmentation of assimilates in the leaf, and their gradual release into the transport pathway, i.e. the input function of the model, may or may not be considered. Also, in some models, the lateral movement of solutes from the sieve cells is taken to be irreversible, whereas others assume a reversible exchange.

By simply choosing parameters and boundary conditions it has proved possible to construct a wide variety of apparently satisfactory models. That is, they can all produce theoretical assimilate and tracer profiles which are similar to those observed. Because of this, it is now necessary to make extensive tests of these models. The tests will have to include comparisons of the behaviour of the model with real situations. In addition, the values of parameters used in the models, for example the hydraulic conductivity of the sieve cells and the permeability of the plasmamembranes to sucrose, will have to be examined critically. If the models do no more than encourage more estimates of factors such as these the effort expended on modelling will probably be worthwhile.

Further reading and references

Further reading

ARNOFF, S., DAINTY, J., GORHAM, P. R., SRIVASTAVA, L. M. and SWANSON, C. A. (1975) *Phloem Transport*. Plenum Press, New York.
CANNY, M. J. (1973) *Phloem Translocation* Cambridge University Press, Cambridge.
MACROBBIE, E. A. C. (1971) Phloem translocation facts and mechanisms: a comparative survey *Biol. Rev.*, **46**, 429–81.
PEEL, A. J. (1974) *Transport of Nutrients in Plants*. Butterworths, London.
WARDLAW, I. F. and PASSIOURA, J. B. (1976) *Transport and Transfer Processes in Plants* Academic Press, New York.
WEATHERLEY, P. E. and JOHNSON, R. P. C. (1968) The form and function of the sieve tube: a problem in reconciliation, *Int. Rev. Cytol.*, **24**, 149–92.
ZIMMERMANN, M. H. and MILBURN, J. A. (eds) (1975) *Transport in Plants I: Phloem Transport, Encyclopedia of Plant Physiology*, Vol. 1 (New Series) Springer-Verlag, New York.

Other references

BIELESKI, R. (1966) Accumulation of phosphate, sulphate and sucrose by excised phloem tissues, *Plant Physiol.*, **41**, 447–54.
DAINTY, J., CROGHAN, P. C. and FENSOM, D. S. (1963) Electro-osmosis, with some applications to plant physiology, *Can. J. Bot.*, **41**, 953–66.
FENSOM, D. S. (1975) *Other Possible Mechanisms*, pp. 354–66, *Transport in Plants I: Phloem Transport, Encyclopedia of Plant Physiology*, Vol. 1 (New Series) Zimmermann M. H. and Milburn, J. A. (eds) Springer-Verlag, Berlin.
FISHER, D. B. (1970) Kinetics of C-14 translocation on soybean. I. Kinetics in the stem, *Plant Physiol.*, **45**, 107–13.
GLASZIOU, K. T. and GAYLOR, K. R. (1972) Storage of sugars in stalks of sugarcane, *Botan. Rev.*, **38**, 471–90.
HALL, S. M. and BAKER, D. A. (1972) The chemical composition of *Ricinus* phloem exudate, *Planta (Berl.)*, **106**, 131–40.

HALL, S. M., BAKER, D. A. and MILBURN, J. A. (1971). Phloem transport of ^{14}C-labelled assimilates in *Ricinus*, *Planta (Berl.)*, **100**, 200–7.

HO, L. C. (1976) The relationship between the rates of carbon transport and of photosynthesis in tomato leaves, *J. exp. Bot.*, **27**, 87–97.

KUO, J., O'BRIEN, T. P. and CANNY, M. J. (1974) Pit-field distribution, plasmodesmatal frequency, and assimilate flux in the mestome sheath cells of wheat leaves, *Planta (Berl.)*, **121**, 97–118.

MALEK, F. and BAKER, D. A. (1977) Proton co-transport of sugars in phloem loading, *Planta (Berl.)*, **135**, 297–9.

MASON, T. G. and MASKELL, S. J. (1928*a*) Studies on the transport of carbohydrates in the cotton plant: I. A study of diurnal variation in the carbohydrates of leaf, bark, and wood, and of the effects of ringing, *Ann. Bot.*, **42**, 189–253.

MASON, T. G. and MASKELL, E. J. (1928*b*) Studies on the transport of carbohydrates in the cotton plant: II. The factors determining the rate and the direction of movement of sugars, *Ann. Bot.*, **42**, 571–636.

MOORBY, J., EBERT, M. and EVANS, N. T. S. (1963) The translocation of ^{11}C-labelled photosynthate in the soybean, *J. exp. Bot.*, **14**, 210–20.

MOORBY, J. and JARMAN, P. D. (1975) The use of compartmental analysis in the study of the movement of carbon through leaves, *Planta (Berl.)* **122**, 155–68.

OUTLAW, W. H. and FISHER, D. B. (1975) Compartmentation in *Vicia faba* leaves: I. Kinetics of ^{14}C in the tissues following pulse labelling, *Plant Physiol.*, **55**, 699–703.

PEARSON, C. J. (1974) Daily changes in carbon dioxide exchange and photosynthate translocation of leaves of *Vicia faba*, *Planta (Berl.)*, **119**, 59–70.

RICHMOND, P. and WARDLAW, I. F. (1976) On the translocation of sugar: van der Waals' forces and surface flow, *Aust. J. Plant Physiol.*, **3**, 545–9.

TAMMES, P. M. L. and VAN DIE, J. (1964) Studies on phloem exudation from *Yucca Blaccida* haw: I. Some observations on the phenomenon of bleeding and the composition, *Acta Botan. Neerl.*, **13**, 76–83.

THAINE, R. (1969) Movement of sugars through plants by cytoplasmic pumping, *Nature*, **222**, 873–5.

TROUGHTON, J. H., MOORBY, J. and CURRIE, B. G. (1974) Investigations of carbon transport in plants: I. The use of Carbon-11 to estimate various parameters of the translocation process, *J. exp. Bot.* **25**, 684–94.

WEATHERLEY, P. E. (1972) Translocation in sieve tubes. Some thoughts on structure and mechanism, *Physiol. Vég.*, **10**, 731–42.

WEATHERLEY, P. E., PEEL, A. J. and HILL, G. P. (1959) The physiology of the sieve tube. Preliminary experiments using aphid mouth parts, *J. exp. Bot.*, **10**, 1–16.

ZIMMERMANN, M. H. (1957) Translocation of organic substances in trees: II. On the translocation mechanism in the phloem of White Ash (*Fraxinus americana* L.), *Plant Physiol.*, **32**, 399–404.

ZIMMERMANN, M. H. (1958) Translocation of organic substances in the phloem of trees, in K. V. Thimann (ed.) *The Physiology of Forest Trees*, pp. 381–400. Ronald Press, New York.

Chapter 3

The movement of water and ions

3.1 Introduction

This chapter will deal with the uptake and transport of water and ions. The mechanisms by which they move across membranes and are transported through cells have been discussed in detail in other volumes in this series – *Water Flow in Plants* by J. A. Milburn and *Cell Membranes and Ion Transport* by J. L. Hall and D. A. Baker. Here, we will be concerned with the general pathways followed by water and ions, the factors which govern the amounts of these substances which enter plants and the interrelationships between their transport. Movement into single cells will be discussed initially and the description then extended to multicellular plants growing in soil.

3.2 The movement of water into cells

Water will move spontaneously between two regions if there is a difference in the free energy of the water in the two regions. The movement is always from the region with the higher (μ, J mol^{-1}) to that with the lower free energy (μ_0). Hence, it is most useful to discuss the movement of water from the soil to the atmosphere in terms of the free energy of water in the various parts of the system i.e. the capacity of the water to do work. The energy available for transport is given, therefore, by:

$$\mu - \mu_0 = \triangle \mu = RT \ln a_w = RT \ln P/P_o \qquad [3.1]$$

where R is the gas constant, 8.31 JK^{-1} mol^{-1}; T is the absolute temperature, K; a_w is the chemical activity of water; P the vapour pressure of water in the system, Pa or Nm^{-2}; and P_o the saturated vapour pressure.

As on any scale of comparison or measurement it is necessary to have a point of reference. In the present context the free energy of water in a situation is always referred to that of pure, free water at the same temperature and pressure. The energy of free water is always taken to be zero. In contrast, the free energy of water in soil or in a cell is always less than that of free water because of the presence of solutes and forces associated with the solid matrix in which the water exists, i.e. the free energy is a negative quantity.

Since we usually express the movement of water in terms of mass, or sometimes volume moved, it has now become standard practice to express the energy of the water not as the free energy, or chemical potential per mole, i.e. J mol^{-1}, but as the water potential, Ψ, with the units of energy per unit mass, J kg^{-1}, where

$$\Psi = \Delta\mu/M_w \qquad\qquad [3.2]$$

and M_w, kg mol^{-1}, is the molar mass of water. Again, water potentials are referred to that of free water and are usually negative.

(It should be noted that in much of the literature water potentials are expressed in units of pressure; usually atmospheres or bars. Neither of these are SI units and will not be used here. However, the relationship between these units and J kg^{-1} can be shown easily. The dimensions of the unit of energy, the joule, are kg m^2 s^{-2}. If then we express water potentials in terms of energy per unit volume the dimensions become kg m^{-2} s^{-2} m^{-3} = kg m^{-1} s^{-2}. These are the units of force per unit area i.e. pressure. Hence the use of atmospheres or bars. NB: 10^2 J kg^{-1} = 1 bar = 0.987 atmospheres = 10^5 Pa = 10^5 Nm^{-2})

The free energy of water is reduced by two types of force. The dissolved solutes in a dilute solution produce what is termed an osmotic potential, π, which is in proportion to the number of solute particles and is independent of the type of particle. Hence,

$$\pi \approx - RTC_s \qquad\qquad [3.3]$$

where C_s is the molal concentration of the solute. The second type are matric forces, τ. These are composed of surface and capillary forces which, for example, hold water in the soil pores or interstices of cell walls and can be described by

$$\tau = - 4\,\sigma/D \approx -2.91/D \ Nm^{-2} \qquad\qquad [3.4]$$

where σ is the surface tension of water (Nm^{-1}) and D (m) is the diameter of the pores in question. The matric forces in vacuolated cells are small relative to the osmotic potential and are often neglected. In soils, meristematic cells and seeds, however, they can become more important than the osmotic potentials.

These negative forces are opposed by positive pressures which can be hydrostatic in origin, for example when comparing water potentials at different depths in the soil or heights in a tree. In cells the positive term is provided by the turgor pressure. The cytoplasm of all living plant cells is surrounded by a semipermeable plasmamembrane. If the cell is placed in water or very dilute solutions the negative osmotic and matric potentials cause water to enter the cell, i.e. it moves into a region of more negative water potential. This uptake of water causes the cell volume to increase and to become less negative because of dilution of the cell contents. Because the cell wall is not completely elastic cell expansion is limited and the continued entry of water results in an increased pressure inside the cell, the turgor pressure, P. The entry of water continues until at equilibrium the water potential, i.e. the sum of the three forces discussed, is equal to the water potential of the external solution, Ψ ext and

$$\Psi = - \pi - \tau + P = \Psi \text{ ext} \qquad [3.5]$$

It should be noted that the movement of water is dependent on a gradient in the free energy of water i.e. the water potential. Matric forces do not have a direct effect on movement; they influence the free energy of the water in the system.

If the external solution is pure water, and the matric potential is neglected, $\pi = P$ when there is no further uptake of water. The cell volume is then maximal and the cell is said to be fully turgid. If the external solution is not pure water $P = \pi - \Psi$ ext. When $P = 0$, i.e. there is no turgor pressure, the cell is said to be at incipient plasmolysis.

The distribution of water in a vacuolated cell, and the relative contribution of π, τ and P to Ψ are shown diagrammatically in Fig. 3.1. In the cell wall matric forces are relatively more important than osmotic forces whereas the reverse is true in the vacuole. It is impossible to distinguish between the two in the case of the cytoplasm. Turgor pressure can only operate inside the plasmamembrane i.e. in the cytoplasm and vacuole.

The movement of water through the plasmamembrane into the cell can be described by an equation analogous to that used in Chapter 2 to describe the entry of CO_2. Here, the appropriate notation is

$$d\,V_i/dt = L_p\,A\ (\Psi - \Psi\text{ ext}) \qquad [3.6]$$

where $V_i\,(m^3)$ is the volume inside the plasmamembrane, $L_p\,(m\,s^{-1}\,J^{-1}\,kg)$ is the permeability of the membrane to water i.e. the hydraulic conductivity, $A\ (m^2)$ is the area of the membrane and $\Psi - \Psi$ ext the gradient of water potential across the membrane. Hence, the rate at which the volume of a cell changes in response to variations in Ψ is a

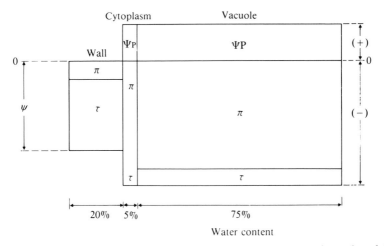

Fig. 3.1 The distribution of water between the cell wall, cytoplasm and vacuole and the relative contributions to the water potential in the three phases of osmotic and matric potentials and turgor pressure (after H. Barrs, private communication).

function of the hydraulic conductivity, L_p. In addition, it will also be affected by the elastic properties of the plasmamembrane and cell wall. Since it is possible that these properties of the cell are pressure sensitive, it can be seen that a completely unambiguous treatment of water uptake by cells would be complex. It is certainly unnecessary for our purposes.

Another, and probably more serious complication, is that the movement of water cannot be considered independently from that of solutes. Again, we need not be concerned with these interactions in detail, but their existence should be remembered. Their operation is discussed in some of the references listed at the end of this chapter and in the book by Hall and Baker in this series.

3.3 The movement of ions into cells

Unicellular plants growing in a dilute solution of salts are usually surrounded by all the mineral nutrients they need. Unfortunately, the relative concentrations of the nutrients in the solution are usually not ideal for the plant which has, therefore, to discriminate between the various ions. Some of the discrimination is exerted on the uptake processes, the entry of some ions being more rapid than others. In addition, there are mechanisms for returning ions to the surrounding solution or accumulating them in vacuoles where they cause less interference with the metabolism of the cell.

The movement of water and ions

The solution in which the cell grows permeates the cell wall up to the plasmamembrane and the latter is the first major barrier to entry. Using the definition given in section 1.4, the solution in the cell wall constitutes the apoplast or free space of the cell. This simple picture is complicated by the presence of, mainly negative changes on the plasmamembrane and carboxyl groups of the wall material. Cations are attracted to these negative changes and form a 'double layer' of electrical charges. Hence, in the regions near the surfaces there is a greater concentration of cations and lower concentrations of anions than in the bulk of the solution (Fig. 3.2).

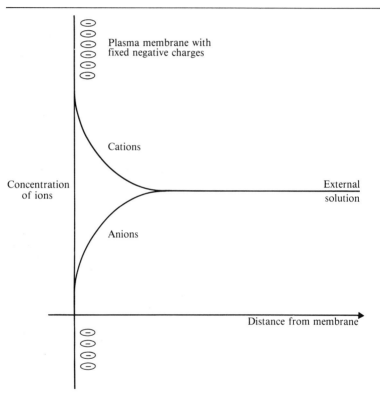

Fig. 3.2 The distribution of anions and cations near a surface carrying fixed negative charges.

If single cells, detached roots, or discs of storage organs such as beet root or potato tuber, are placed in a solution containing a radioactive ion as a tracer, the uptake of the ion can be followed easily. The results of such an experiment are shown diagrammatically in Fig. 3.3. It can be

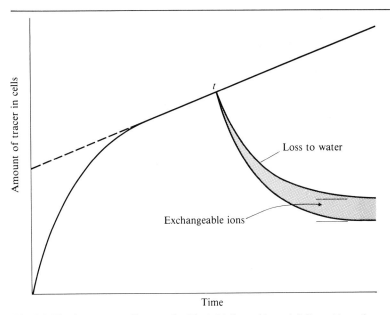

Fig. 3.3 The time course of ion uptake. The initially rapid rate is followed by a slower phase. Extrapolation of the slower phase to time zero gives an estimate of the ions in the apoplasm. When the tissue is transferred to water some of the ions are rapidly leached from the tissue. A rather larger proportion of the absorbed ions is removed by an exchange solution.

seen that there is an initially rapid rate of uptake followed by a slower phase. If at time t the cells are transferred to a similar, but unlabelled, solution, or water, some of the tracer is lost from the cells. A rather larger fraction is lost to the unlabelled solution than to water because the unlabelled ions exchange with the tracer ions absorbed on the charged sites in the apoplast. The tracer ions which remain in the cells after the period of exchange are thought to have moved across the plasmamembrane into the symplast.

Although the concentration of many ions is greater inside the cell than in the surrounding solution it does not necessarily mean that these ions have been pumped into the cell by means of some metabolically powered pump. By definition, ions carry electrical charges and hence their movement across a membrane will be aided or hindered by the presence of a potential difference across the membrane. Such a potential difference can be generated and maintained by a variety of factors; the diffusion of ions in response to a concentration gradient, the nature of the membrane and differing rates of metabolism on the two sides of the membrane.

At equilibrium, diffusion in response to concentration differences is balanced by movement in response to electrical effects and the situation is described by the Nernst equation:

$$Z_j F (P_i - P_o) = RT \ln (a_{jo}/a_{ji}) \qquad [3.7]$$

where Z_j is the valency of the ion j (positive or negative); F (C mol^{-1}) is Faradays' constant; $P_i - P_o$ (mV) is the electrical potential gradient; and a_{ji}/a_{jo} is the ratio of the chemical activity of ion j inside and outside the membrane ($a_j = \gamma_j C_j$ where γ_j is the activity coefficient and C_j the concentration (mol) of ion j). Modest potential differences can support considerable concentration differences. Substitution in equation [3.7] shows that a potential inside a membrane of -116 mV relative to the outside could maintain a 100-fold increase in concentration of a cation without the expenditure of any metabolic energy. Metabolic pumps need only be invoked when the concentrations are greater than those which can be maintained by the electrochemical potential gradient. In such instances, the energy may come from photosynthetic phosphorylation in illuminated cells or from respiration. The various theories of how the energy is used to move ions across membranes are covered in detail in Hall and Baker in this series and need not be repeated here.

3.4 The movement of water and ions into the xylem

The forces which govern the movement of water into unicellular plants, i.e. gradients in water potential, also produce the movement of water from the soil, through the plant and into the atmosphere. Some approximate values of water potentials in the system are shown in Table 3.1.

As described in section 3.3 the walls of cells are accessible to the bathing solution. Because there is no barrier to the movement of the solution throughout the interconnecting cell walls of the cortical and epidermal cells it follows that the apoplast of a root extends throughout the walls of these tissues.

Table 3.1 Approximate ranges of values of the water potential (J kg^{-1}) in the soil, leaves and atmosphere (after J. L. Milthorpe, and J. Moorby, 1974).

	Turgid plant	Wilting plant
Soil	-10 to -1000	-1000 to -2000
Leaves	-200 to -1500	-1500 to -3000
Atmosphere	$-10\,000$ to $-200\,000$	$-10\,000$ to $-2\,000\,000$

Water and ions will enter the cells which border the apoplast in the ways set out in section 3.2 and 3.3, but this cannot be considered to be a complete description of their uptake by a root. A root is not an amorphous mass of cells; it has a definite structure which operates as a system for gathering materials from the soil and transporting them to the xylem. This may be demonstrated by the phenomenon of root pressure. Thus, if the stem is removed from a root system, liquid is exuded from the cut ends of the xylem vessels in the root. This liquid can be collected, and analysis shows it to be a solution containing the same ions as those present in the solution surrounding the roots, but they are usually more concentrated and are present in different relative amounts.

The whole root appears to act as an osmometer, with water moving into the xylem down a water potential gradient. It is obvious that the gradient is caused by the greater concentration of ions in the xylem sap than in the surrounding solution, but it is less obvious how this increase in concentration is produced. There must, however, be one or more membranes somewhere in the pathway to the xylem in order for the concentration differences to be maintained.

If the concentration of the ions surrounding the roots is increased there are corresponding increases in the fluxes of water and ions into the xylem and the concentration of the ions in the xylem sap (Fig. 3.4). These results show that the effects described only occur when the concentration of the external solution was less than about 1 m mol KCl; concentrations greater than this had no further effect. When the data at concentrations up to 1 m mol were plotted on an expanded scale (Fig. 3.4b) they showed there was still movement of water and K into the xylem when there was no K in the external solution. This K must have been absorbed before the start of the experiment and moved out of the other root cells when the external concentration of K was reduced to zero.

These results have been interpreted in terms of two fluxes of K, one dependent, and the other independent, of the flux of water. In this experiment it was suggested that they were partitioned in the way shown in Fig. 3.5. Other experiments have also shown interactions between the external concentration of ions and the rate of transpiration on the rate of transfer of ions to the xylem. The mineral nutrient status of the plants is also an important factor. The situation appears to be that if plants are adequately supplied with nutrients the concentration of nutrients in the xylem is similar to that in the external solution and is unaffected by the rate of transpiration. In contrast, if the nutrient status of the plants is low the concentration of ions in the xylem tends to be greater than in the external solution.

This interpretation does not provide any information on how the increase in ionic concentration is achieved. Other experiments have

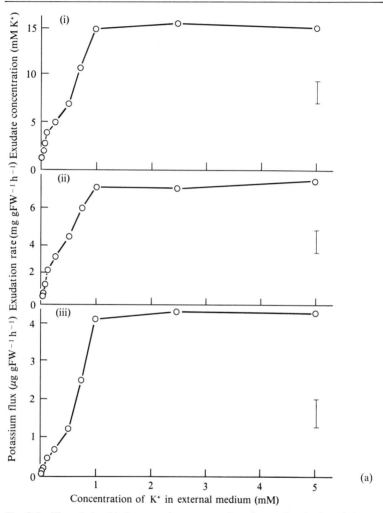

Fig. 3.4a The relationship between the concentration of potassium in the solution around the roots of a *Ricinus* plant and the (i) the concentration of potassium in the exudate from the detopped plant; (ii) the rate of exudation and (iii) the flux of potassium in the exudate. The vertical lines are pooled standard deviations.

shown that the transport of, for example, K can usually be ascribed to the electrochemical potential gradient, i.e. it is usually passive. In contrast, the transport of Cl is often active, and requires the expenditure of metabolic energy.

In the preceding discussion water uptake was considered only as a

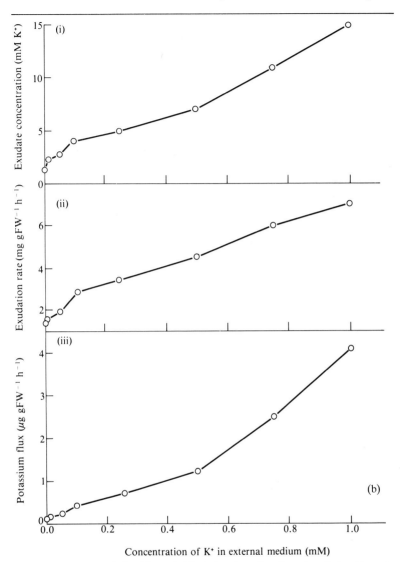

Fig. 3.4*b* The same relationships as in Fig. 3.4*a* enlarged to show the relationship up to an external concentration of 1.0 mM K+ (after D. A. Baker and P. E. Weatherley, 1969).

function of the osmotic gradient across the root cortex. This ignores the effect of transpiration. It is relevent, therefore, at this point to consider the forces acting on water uptake and transport through the

85

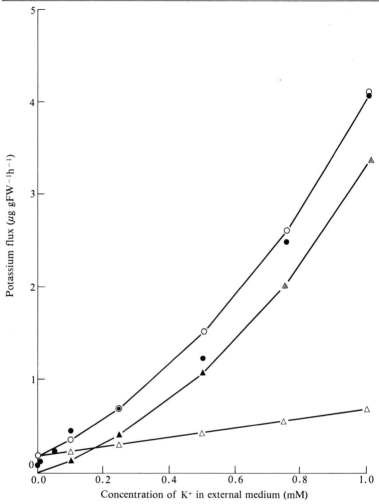

Fig. 3.5 The relationship between the external K^+ concentration and the total observed K^+ flux in the exudate from a detopped *Ricinus* plant ($\bullet$); the calculated water-dependent flux ($\triangle$); the calculated water-independent flux ($\blacktriangle$); and the sum of the two calculated fluxes ($\bigcirc$) (after D. A. Baker and P. E. Weatherley, 1969).

intact plant. A method often used to describe this system is analogous to that used in section 2.1 to describe the movement of carbon into leaves. There, the rate of CO_2 uptake was proportional to the driving force, i.e. the difference in CO_2 concentration, and inversely proportional to a transport resistance. In a similar fashion the flow of water through the plant is proportional to the difference in water potential between the two ends of the system and inversely proportional to the resistance to transport of water between the ends. Hence,

$$F = \triangle \Psi / R'$$ [3.8]

The pathway from the soil to the atmosphere can be divided into sections and the movement through each section considered in relation to the relevant gradients in water potential and resistances. However, it is difficult to make accurate estimates of flow and resistance in this way because of the ability of the plant to call on considerable quantities of water stored in the various tissues. In some experiments with potatoes, for example, uptake of water by the roots was unable to satisfy the water lost by transpiration. As a result, water moved out of the tubers and there was a 10 per cent decrease in tuber fresh weight (Fig. 4.19). Similarly, there is often a reduction in the circumference of stems and the volume of fleshy fruits in periods of rapid transpiration.

The overall water conducting system can, therefore, be considered as a resistance network with associated capacitors (Fig. 3.6). These latter provide some buffering and can cause time lags in the responses of individual parts of the system to changes in other regions. For example, the rate of water uptake usually lags behind the rate of loss of water from a plant. It is this time difference which can lead to the production of water deficits in a plant even when the roots have ample supplies of water.

There are differences in the relative magnitudes of the resistances to water transport in the various parts of the plant. Experiments in which roots have been removed from transpiring plants whilst maintaining the supply of water to the stem have often shown an increase in the rate of transpiration. This suggests that the resistance to flow in the roots is greater than the resistances in the stem and leaves. Direct estimates tend to confirm this conclusion, the foliage and stem resistances usually being approximately equal and the root resistance 2–4 times larger.

One explanation of this effect is that it is a consequence of the structure of the root. Movement of both water and ions through the epidermis and cortex probably takes place primarily through the apoplast. In the very young tissues near the root apex the apoplast probably extends through to the xylem. In older tissues, however, movement beyond the endodermis is prevented by the development of the Casparian strip in the anticlinal walls of the endodermal cells (Fig. 3.7). The Casparian strip is formed by the deposition of a band of suberin and lignin in the cell walls and appears to reduce the permeability of the walls to water and ions since it forms a barrier on all but the tangential walls. It effectively 'waterproofs' the tissues within

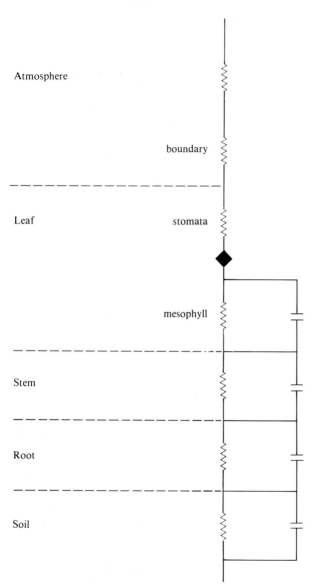

Fig. 3.6 The pathway followed by water through the plant from the soil to the atmosphere layed out as an electrical circuit. The diamond between the mesophyll and the stomata indicates that there is a change of phase at this point from liquid water to water vapour.

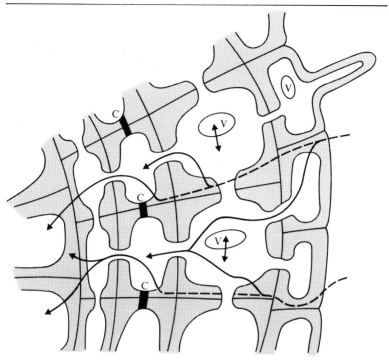

Fig. 3.7 A simplified diagram showing the pathways followed by water and ions as they move across the root to the xylem. Dotted lines indicate movement in the apoplast and solid lines symplastic movement. The shaded areas are cell walls, the regions marked V, vacuoles, and C is the Casparian strip.

the endodermis. This change is enhanced in some species by the laying down of suberin lamellae on the internal walls of the endodermis.

The physiological implications of these structural changes is that they ensure that materials only enter the xylem after movement through the symplasm of the plant. The plant can, therefore, exert some control over this movement and the implications of this control for ion transport are discussed below.

The movement of water through the endodermal cells is likely to be the source of the relatively greater resistance to water transport through the root than through the stem where, as will be seen in section 3.5, transport is mainly through the apoplast. It should be appreciated that the Casparian strip does not form a complete barrier. There are breaks, especially in older tissues through which transport can occur in the apoplast, and water uptake can take place through very suberized woody roots.

It should be noticed that two types of force act to change the gradient in water potential across the root. The suction in the xylem cells will tend to cause a pressure drop, and the difference in solution concentration between the xylem sap and the external solution will lead to an osmotic gradient. It is possible to manipulate these two types of force independently and it is usually found that the resistance to flow induced by a pressure difference is less than the resistance when the water flow is induced osmotically. As explained earlier (cf. equation [2.4]), the flow of a liquid through a channel in response to a pressure difference is proportional to the fourth power of the radius of the channel. Therefore, as the pressure difference across a root is increased it can be envisaged that flow takes place through larger numbers of progressively smaller channels. This type of response is unlikely, however, when the flow is generated osmotically.

There are yet further complications to the pattern of water uptake and transport. It has often been found that the root resistance declines as the rate of transpiration, i.e. the flux of water through the plant, is increased. In addition, the results of some experiments have shown diurnal fluctuations in the root resistance. Both of these effects probably arise from changes in the factors which influence uptake into the symplasm and the relative importance of water transport through the symplasm and apoplasm. For example, there are diurnal fluctuations in the rate of ion uptake by plants (Fig. 4.1) which, in turn, will affect the osmotic component of the water flux. Further, as the flux of water through a plant increases, the water is likely to take the pathways of least resistance. Hence, relatively more of the water will tend to move through the apoplasm and so the overall resistance will decrease.

The restrictions placed on the movement of water across the root and into the xylem are also faced by the ions which enter the root. Near the root apex the absence of a Casparian strip allows ions to move into the xylem through the apoplasm. Once the Casparian strip is formed, transfer across the endodermis has to take place in the symplasm. The permeability of the plasmamembrane, and the transport processes acting there, allow the plant to exert some control over the passage of ions into the xylem. In addition, the movement of ions across the tonoplasts into and out of the vacuoles, and the possibility of the secretion of ions into the xylem from the xylem parenchyma cells, provide further possibilities for control.

Acceptance of this description demands that there are sufficient plasmodesmata of adequate size to maintain the necessary fluxes. Calculations suggest that this might well be the case. In some experiments the flux of phosphate across the inner tangential wall of the endodermis of barley roots was 3.8×10^{-9} mol m^{-2} s^{-1} and the flux through a single plasmodesmata was 3.6×10^{-21} mol s^{-1}. A similar estimate of the flux of potassium through the endodermis of marrow was 140×10^{-9} mol m^{-2} s^{-1} and that through a plasmodesmata 2.19×10^{-20} mol s^{-1}. The phosphate fluxes calculated for barley are less than estimates made using other systems, but the potassium fluxes in marrow are rather greater.

One way of investigating the physiological consequences of the development of the Casparian strip is to compare its development, as a root extends, with the patterns of ion uptake along the axis of the root. It is possible to perform this type of experiment using equipment shown diagrammatically in Fig. 3.8. Knowing the specific activity (Ci g^{-1}) of the labelled solution measurements of the amount of tracer permit the amount of a nutrient absorbed or moved out of the region exposed to the label to be calculated.

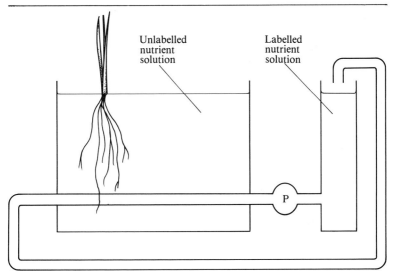

Fig. 3.8 The type of equipment used to monitor ion uptake and transport from a small part of a root. The plant is placed in unlabelled, aerated nutrient solution and a root is sealed into a hole or slit made in tubing. Labelled nutrient solution is circulated through the tubing from a reservoir by a pump, P.

Results obtained in this way can be seen in Fig. 3.9. Although the greatest rate of uptake of both phosphate and calcium was near the root apex, the uptake of both ions proceeded at a substantial rate along the lengths of all the root types. The rate of phosphate uptake showed a progressive decline through the sequence seminal, nodal and primary lateral roots. This effect was not obvious with calcium uptake, the rate of which tended to be about half that of phosphate uptake. A major difference in the results is that although there is a considerable amount

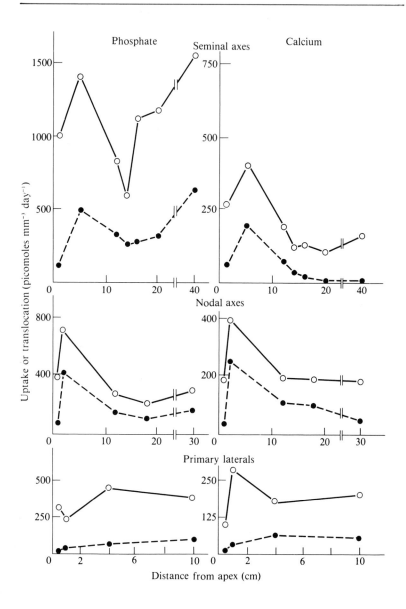

Fig. 3.9 The uptake (○——○) and transport (●----●) of phosphate and calcium in 24 h by 3.5 mm segments of various types of *Hordeum vulgare* roots. This data was obtained using equipment similar to that shown in Fig. 3.9 (after D. T. Clarkson and J. Sanderson, 1971).

of phosphate transport from all the root zones there was little transport of calcium from the older, more proximal regions of the seminal roots. Similar results, plotted in a different way to emphasize these points, are shown in Fig. 3.10.

When these results are compared with anatomical studies, it is found that the transport of calcium is least in those regions where there is extensive development of the Casparian strip. Other experiments suggest that the uptake and transport of iron follows a similar pattern. In contrast, the movement of potassium and phosphate do not seem to be affected to any great extent by the development of the Casparian strip. These ions appear to pass readily through the symplasm of the xylem whereas the symplastic transport of calcium and iron is less effective.

If the type of data shown in Fig. 3.9 are combined with estimates of the amounts of the various categories of roots in the total root system, it is possible to calculate the contributions made by the different root members to the total amount of nutrients absorbed by the plant (Table 3.2). These data suggest that the lateral roots absorbed almost half of the phosphate. However, because only a small proportion of the ions

Table 3.2 Calculated amounts of phosphate and calcium absorbed by the different members of the root system of a 4-week-old barley plant expressed as the percentages of the total amounts absorbed or translocated. The amounts in parenthesis show the amounts translocated as percentages of the amounts absorbed by the individual root members (after D. T. Clarkson, and J. Sanderson, 1970).

Root member	Phosphate		Calcium	
	Uptake	Transport	Uptake	Transport
Seminal	26	34 (33)	12	6 (22)
Early nodal	7	14 (48)	17	17 (40)
Late nodal	22	32 (38)	42	61 (57)
Laterals	45	19 (11)	29	16 (21)

they absorbed were transported to the rest of the plant the laterals contributed only about the same proportion as the early nodal roots to the total supplied to the shoots. Most of the calcium was absorbed by the late-formed nodal roots, which also transported a large proportion of these ions to the rest of the plants.

The relationships discussed above were determined using small plants growing in nutrient solutions. Different situations exist with soil-grown plants. The concentrations of ions around the roots vary both spatially throughout the soil profile and with time. There will, therefore, be a direct effect of these concentrations on the rates of ion uptake. In addition, there will be an indirect effect. High

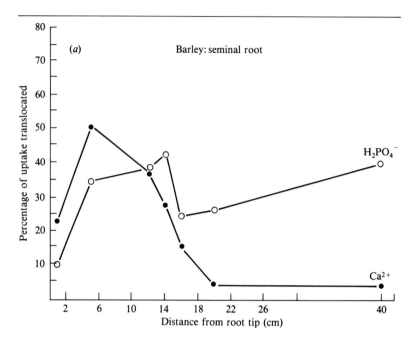

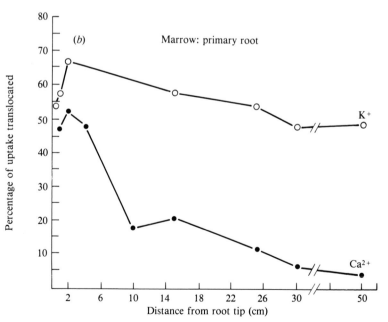

◀ **Fig. 3.10** Percentages of Ca^{++} (●——●) or $H_2PO_4^-$ ions (○——○) absorbed by (*a*) *Hordeum vulgare* and (*b*) *Cucurbita pepo* roots which were transported to the aerial parts of the plants. (Copyright © 1974 McGraw-Hill Book Co. (UK) Limited. From Clarkson: *Ion Transport and Cell Structure in Plants*. Reproduced by permission of the author and publisher.)

concentrations of some ions, e.g. nitrate, can cause increased branching of the root system. Hence, localized differences in the concentrations of these ions can lead to increases in the size of the root system in these regions.

3.5 The movement of water and ions through the xylem

The major factor governing the movement of water and ions through the xylem and into the leaves is the rate of transpiration. The evaporation of water into the intercellular spaces of the leaf reduces the amount of water in the cells. This, in turn, results in a reduced turgor pressure, i.e. increased suction, in the xylem cells of the leaf veins. The water columns in the xylem are, therefore, in a state of tension and water is pulled into the leaves from the remainder of the plant.

The tensile strength of the water columns in the xylem cells is sufficient to withstand the tensions necessary to raise the water through the plant, even through tall trees, if the columns remain intact. Once they are broken, however, and contain air bubbles, they cannot function until the embolisms have been removed. This is a real problem in trees exposed to low temperatures because the freezing of the xylem sap can cause fragmentation of the columns. There is some evidence that, in the long term, the air bubbles may be redissolved in the xylem sap. Alternatively, root pressures, which are often of the order of 1 to 2×10^5 Nm^{-2} could refill empty xylem cells up to heights of 10 to 20 m.

In many species the dimensions of the pits are so small that when an embolism does occur surface tension forces prevent the passage of the gas through the pits, so restricting the spread of the bubble. Where bordered pits are present, the overall dimensions of the pit opening are larger, but the pores in the margo of the pit membrane, through which transport takes place, are too small to allow the passage of the gas bubble. The differences in pressure on the two sides of the pit then cause the movement of the torus against the shoulders of the pit so sealing the opening.

Gravity imposes a gradient of increasing pressure with height equivalent to 10^4 Nm^{-2} m^{-1}. Any movement of solution up a plant

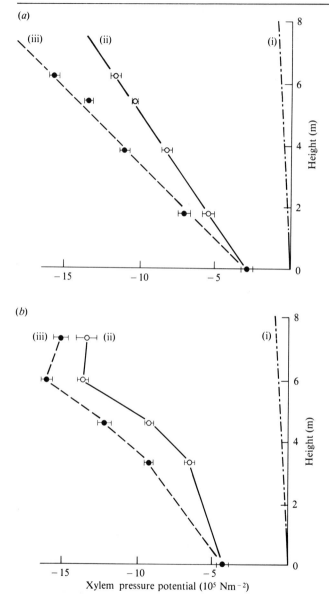

Fig. 3.11 The gradients in xylem pressure potential in *Picea* stems (*a*) on a dull cool day (*b*) on a sunny warm day. (i) A gradient of $10^4\ Nm^{-2}m^{-1}$; that to be expected from the effect of gravity alone (–·–·–) (ii) The potentials in the trunk of the tree (○——○). (iii) The potentials in the terminal shoots of branches inserted at the same height (●----●). All ± standard errors (after J. Hellkvist, G. P. Richards and P. G. Jarvis, 1974).

through the xylem will necessitate, therefore, a water potential gradient, i.e. suction gradient, in excess of this value. Measured gradients in transpiring trees not suffering from a water stress are often about 50 per cent greater than this value and can be up to about 2×10^5 $Nm^{-2} m^{-1}$ (Fig. 3.11). At equivalent transpiration rates the presence of the larger gradients in water potential would indicate a greater resistance to movement through the xylem (equation [3.8]). It has been suggested that some of this variation can be attributed to interspecific differences in the dimensions of xylem cells.

As explained in the previous section, some of the water evaporated from the transpiring surfaces can come from outside the direct transpiration pathway. The xylem is composed of dead cells outside the plasmamembrane and, as such, is part of the apoplast. Some of the water which can be called upon when the transpiration rate is high will be outside the xylem but still in the apoplast; for example in the cell walls of the surrounding cells. Yet other water will be symplastic in origin and has, therefore, to move across a membrane before it can enter the transpiration stream. As in the root tissues, the relative amounts of movement between and through the symplast and apoplast will depend on the water potential gradients and relevant transport

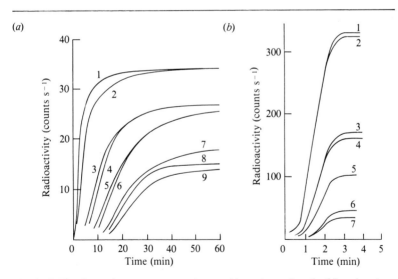

Fig. 3.12 The change in count rate at various positions along a length of *Populus nigra* var. *italica* stem after the application of a labelled solution to the basal end and suction to the apical end. (*a*) 100 cm length of stem, 10^5 N m^{-2} suction, ^{32}P supplied to the basal end and counter 1 to 9 at increasing distances from 7.5 to 50.5 cm from the base respectively. (*b*) 58 cm length of stem, 5.3×10^4 N m^{-2} suction, ^{42}K supplied to the base of the stem. Counters 1 and 2 were 19 cm from the base and 4 at 28 cm, 5 at 35 cm and 6 and 7 at 45 cm respectively (after R. W. Heine, 1970).

resistances. There will, however, be a tendency to equilibration between the two bodies of water in any one region.

Similar arguments to these can be applied also to the movement of ions, although there is not as large a sink for ions in the leaves as is provided for water by transpiration. Hence, recently absorbed ions need not be transported directly to the leaves. Ions can be transferred to other tissues, including the phloem, and this latter can lead to a circulation of ions throughout the plant.

The concentration of ions at any point in the xylem will be the sum of the movement between these various phases. For example, as nutrient shortage or deficiency develops in the plant the amounts of ions transported to the aerial parts decline because the roots tend to retain a greater proportion of the smaller amounts absorbed.

The speeds at which water and ions move through the xylem can be estimated using a heat-pulse technique for water and radioactive tracers for ions. The former method involves fixing small thermocouples, or thermistors, to the stem and using these to detect changes in the stem temperature when a distant part is heated. The

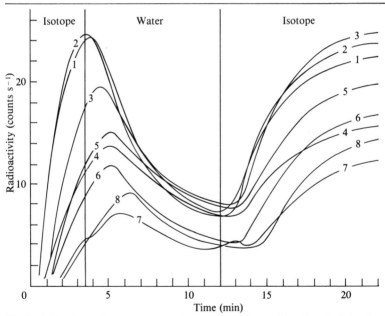

Fig. 3.13 The change in count rate at various positions along a 66 cm length of *Populus nigra* var. *italica* stem. Suction was applied to the apical end of the stem and ³²P was supplied to the base of the stem for 3.5 min followed by water for 8.5 min and then ³²P for the remainder of the experiment. Counters 1 and 2 were 6 cm from the base of the stem, 3 and 4 at 11 cm, 5 and 6 at 16 cm and 7 and 8 at 29 cm respectively (after R. W. Heine, 1970)

heat is conducted mainly through the xylem and any temperature changes will be functions of the rate of sap movement through the xylem and the distances between the thermocouples and the heat source. Labelled ions can be introduced into the xylem either through the roots or through the cut bases of stems. Both techniques are analogous to those which are used to estimate the speed of transport through the phloem and which were discussed in Section 2. All three suffer from the same problems, the most important of which is the relative extents of lateral and longitudinal movement.

The importance of this effect seems to vary, and take different forms, with different ions. For example, the movement of labelled phosphate and potassium ions through the xylem of sections of poplar stems in response to applied pressure is shown in Fig. 3.12. These plots of the arrival of the tracers at increasing distances along the stem are very similar to those showing carbon movement through the phloem (Fig. 2.10a). When the supply of labelled phosphate to the bottom of the stem was replaced by water the amount of ^{32}P in the vicinity of the counters declined markedly (Fig. 3.13). This suggests that it is

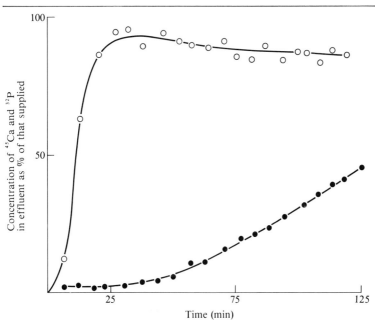

Fig. 3.14 The concentration of ^{45}Ca (●——●) and ^{32}P (○——○) in the exudate from a 12 cm length of *Malus* stem (cv. Granny Smith) after the application of the isotope to the base of the stem under a pressure of $3.45 \times 10^3 \, \mathrm{Nm^{-2}}$. The concentration of the isotope in the exudate is expressed as a percentage of the concentration in the solution supplied to the base of the stem (after J. B. Ferguson and E. G. Bollard, 1976).

relatively easy to flush out the phosphate from the xylem.

In contrast, it proved more difficult to wash out labelled-potassium with water and the behaviour of calcium and other alkaline earth cations seems to be very different from both phosphate and potassium. For example, in analagous experiments to those shown in Figs 3.12 and 3.13 a direct comparison of the movement of ^{45}Ca and ^{32}P through apple wood showed that much less ^{45}Ca was transported (Fig. 3.14).

In other experiments, attempts to wash out the ^{45}Ca with unlabelled nutrient solution or 5 mM KCl were unsuccesful. In contrast, when the labelled nutrient solution was replaced with unlabelled 10 mM CaCl$_2$ the amount of label in the stem declined (Fig. 3.15a). When the plants were transferred to unlabelled solutions of CaCl$_2$ after exposure to the ^{45}Ca for only a short period it was possible to flush out a large proportion of the ^{45}Ca from the xylem. The rate of ^{45}Ca efflux declined as the period of exposure to the tracer was increased (Fig. 3.15b).

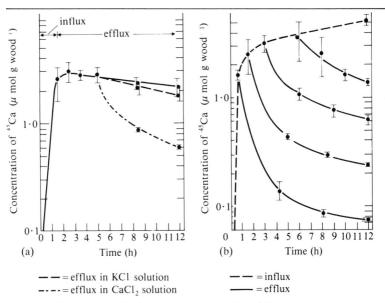

— — = efflux in KCl solution
– – · – = efflux in CaCl$_2$ solution

— — = influx
——— = efflux

Fig. 3.15a The change with time in the concentration of ^{45}Ca in a 2.5 cm section of *Phaseolus vulgaris* stem between the cotyledonary and primary leaf nodes after supplying ^{45}CaCl$_2$ to the roots for 1.5 h 1/100 strength nutrient solution for the next 3.5 h and nutrient (———), 2.5 mM Ca Cl$_2$ (– – · – –) or 5 mM KCl (– · – · –) for the remainder of the experiment. The vertical lines are standard deviations.

Fig. 3.15b The time course of the total ^{45}Ca influx into the stem segment from ^{45}Ca-labelled nutrient solution (– – – –). After varying periods of uptake the plants were transferred to unlabelled nutrient solution and the efflux from the stem sections followed (———) (after C. W. Bell and D. Biddulph, 1963).

These results suggest that the calcium was absorbed onto exchange sites in the xylem and moved up these by a series of exchange reactions. Attempts to prevent such absorption by complexing the Ca with EDTA increased the rate of Ca movement (Fig. 3.16). Citrate also has some complexing action and when it was added to $^{45}CaCl_2$ solution

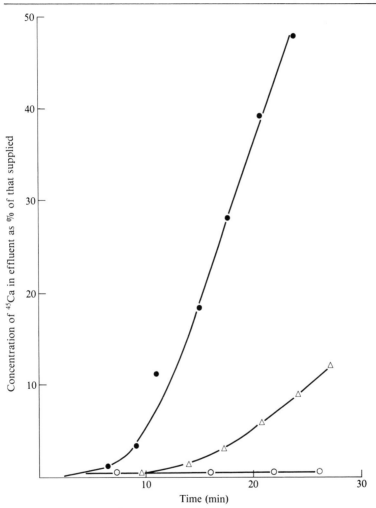

Fig. 3.16 The amount of ^{45}Ca in the exudate from a 12 cm length of *Malus* stem when EDTA was added to the solution supplied to the base of the stem under a pressure of $3.45 \times 10^3 \, Nm^{-2}$. The data are expressed in the same way as in Fig. 3.14. 2 mM $^{45}CaCl_2$ + 0.2 mM EDTA (o——o); 2 mM $^{45}CaCl_2$ + 2 mM EDTA (△——△); 2 mM $^{45}CaCl_2$ + 6 mM EDTA (●——●) (after J. B. Ferguson and E. G. Bollard, 1976).

supplied to an apple stem it also increased the rate of ^{45}Ca movement.

As discussed in section 3.4, the resistance to water flow into the xylem appears to be an inverse function of the rate of transpiration and a similar relationship also appears to hold for the transport of ions through the xylem. For example, it was possible in some experiments similar to those shown in Figs 3.12 and 3.13 to measure both the speed of isotope movement and the rate of exudation of sap from the apical end of the section of stem. Both of these showed a curvilinear relationship with the pressure applied to the basal end of the stem (Fig. 3.17). This would be expected if the resistance decreased as the pressure applied increased.

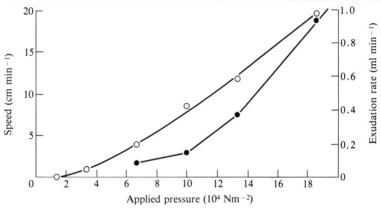

Fig. 3.17 The effect of applied pressure on the speed of ^{32}P movement (●——●) and the rate of exudation (○——○) from stems of *Populus nigra* var. *italica* (after T. M. Morrison and R. W. Heine, 1966).

3.6 The movement of water and ions into and through leaves

The resistance to movement through the apoplast is often less than that to movement through the symplasm. The major pathway followed by the transpiration stream from the xylem to the substomatal cavities is, therefore, probably through the cell walls of the mesophyll cells. This could be seen in experiments in which chelated lead was used as a tracer and was shown to be localized in the cell walls. This apoplastic pathway can be supplemented or reduced by water leaving or entering the mesophyll cells.

The transpiration stream is not pure water, it contains ions which are also delivered to the cell walls of the leaf cells. Most of these ions are absorbed by the cells of the leaf. The efficiency of these absorption processes can be seen in the results shown in Fig. 3.18. The osmotic potential of the exudate from cut petioles and the guttation fluid exuding from the leaf lamina was lower than the osmotic potential of the nutrient solution forced into the petioles. In particular, the osmotic potential of the guttation fluid was reduced to a uniformly low level irrespective of the potential of the solution introduced into the petiole.

There are obvious similarities between the supply of ions to roots and to leaves. In both instances the initial movement is in the apoplast. In these circumstances it is not surprising that there are also similarities in the uptake processes into the symplasm. Work with leaf cells of terrestrial higher plants usually necessitates the use of tissue slices and experimentation is made difficult by the presence of a large proportion

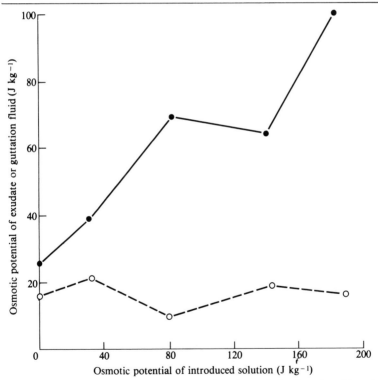

Fig. 3.18 The effect of the osmotic potential of nutrient solution introduced into the petioles of *Capsicum frutescens* leaves under a pressure of 2×10^4 Nm^{-2} on the osmotic potentials of exudate from the cut distal ends of the petioles (●——●) and the 'guttation fluid' exuded from the leaf margins (○----○) (after B. Klepper and M. R. Kaufmann, 1966).

of damaged cells. Nevertheless, in some species, the uptake of some ions has been shown to be an active process with the energy coming from cyclic photophosphorylation in the light and oxidative phosphorylation in the dark. In yet other experiments there appears to be competition between similar ions, e.g. K and Rb, for common uptake sites or pumps.

These remarks have been concerned mainly with the uptake of ions by leaf cells after their delivery in the xylem. In aquatic plants there is the possibility of the absorption of ions by leaves directly from the external solution. This can be an important source of ions to the plant.

3.7 The movement of ions in the phloem

The continued supply of ions to leaves in the transpiration stream suggests that the concentration of these ions should increase through the life of the leaves. However, this increase in concentration occurs with only a few ions indicating that the plant can move ions out of the leaves.

Some plants get rid of ions by transferring them to salt glands. Many halophytes, for example, have glands which secrete both water and excess ions; the latter forming crystalline deposits in dry habitats. Another type of gland allows the plants to dispose of excess salts without losing much water. Here, specific hair cells accumulate the salts plus some water. As accumulation proceeds, the cells swell and eventually burst and die, thus disposing of the salts.

In addition to this removal from the plant, some ions move into the phloem and are transported to other parts of the plant to be used by younger growing organs. The entry of recently absorbed ions into the sieve cells can be directly from the apoplast, or via transfer cells. The ease with which ions enter the sieve cells varies with the age of the leaf and the ion in question. Potassium and phosphate ions, and some nitrogenous compounds, enter the sieve tubes easily. Sulphate, and many micronutrients, enter less readily and in some instances calcium and the other alkaline earth cations appear to be excluded.

Autoradiographic and other evidence suggests that the alkaline earths accumulate in bundle sheath and mesophyll cells in preference to the phloem; only when these sites were swamped by large amounts of unlabelled carrier calcium could ^{45}Ca be induced to enter the phloem. This lack of movement into the phloem can be seen in Fig. 3.19 where ^{89}Sr sprayed onto a potato leaf appeared to move away from the veins and accumulated around the margins of the leaf. In

Fig. 3.19 Autoradiographs showing the distribution of (*a*) ^{89}Sr and (*b*) ^{137}Cs in a leaf ▶ 24 h after potato plants had been sprayed with carrier-free solutions of the isotopes.

(a)

(b)

contrast, ^{137}Cs moved into the veins and was transported to the rest of the plant. These results are compatible with a common syndrome of effects associated with the apparent immobility of calcium but are at variance with the presence of appreciable amounts of calcium in phloem exudates (Tables 2.5 and 2.7).

Once ions and exotic substances such as pesticides have entered the sieve cells they seem to move quite readily along similar pathways to the carbohydrates. Hence, when $^{14}CO_2$ and ^{32}P were supplied to wheat leaves, similar proportions of both tracers moved to the ear and to the basal parts of the plants (Table 3.3). Further, when the upper parts of the plants were shaded to induce increased movement of assimilates to the ear there was a similar change in the distribution of ^{32}P.

Table 3.3 The distribution of ^{14}C and ^{32}P expressed as percentages of the total amount of isotope present 24 h after their application to the leaf below the flag leaf of wheat plants. The results are the mean of six replicates ± standard errors (after C. Marshall and I. F. Wardlaw, 1973).

Plant parts	Controls		Shaded plants*	
	^{14}C	^{32}P	^{14}C	^{32}P
Ear	26.6 ± 4.0	31.5 ± 3.2	89.9 ± 2.3	88.4 ± 2.2
Top and second internodes	6.6 ± 0.6	5.3 ± 0.4	4.0 ± 0.2	5.1 ± 0.4
Flag leaf	1.7 ± 0.2	2.3 ± 0.4	0.1 ± 0.0	0.1 ± 0.0
Third internode	9.0 ± 1.4	8.4 ± 1.3	1.2 ± 0.2	1.6 ± 0.2
Lower stem, roots and crown	56.1 ± 3.3	52.5 ± 2.6	4.9 ± 2.2	5.0 ± 2.1
Relative total activity (counts/min)	651 ± 24	2455 ± 342	1180 ± 90	4275 ± 245

* The sheath and lamina of the flag leaf and the top internode were shaded with aluminium foil.

This is not to say that all materials move to the same extent. The relative rates of exchange between the sieve tube contents and the surrounding tissues play a large part in determining the amounts of materials moved and their speeds of movement.

The movement of ions through the phloem is a most important source of ions for regions of the plant which have limited rates of transpiration. These include, for example, primordia in the apical meristem of shoots, fruit with thick cuticles and storage organs which develop underground. The interrelations between transport and the growth of some of these organs will be discussed in Chapter 4.

Further reading and references

Further reading

BAKER, D. A. and HALL, J. L. (1975) *Ion Transport in Plant Cells and Tissues*. Elsevier, Amsterdam.

CLARKSON, D. T. (1974) *Ion Transport and Cell Structure in Plants*. McGraw-Hill, London.

DAINTY, J. (1969) The water relations of plants, in *The Physiology of Plant Growth and Development*, pp. 421–52, Wilkins, M. B. (ed.), McGraw-Hill, London.

DAINTY, J. (1969) The ionic relations of plants, in *The Physiology of Plant Growth and Development*, pp. 455–85, Wilkins, M. B. (ed.), McGraw-Hill, London.

EPSTEIN, E. (1972) *Mineral Nutrition of Plants: Principles and Perspectives*. John Wiley, New York.

JARVIS, P. G. (1975) *Water Transfer in Plants in Heat and Mass Transfer in the Biosphere*, pp. 369–94, De Vries, D. A. and Algan, N. H. (eds), John Wiley, New York.

LÜTTGE, U. and PITMAN, M. G. (ed.) (1976) *Transport in Plants II: Part A and Part B. Encyclopedia of Plant Physiology* Vol. 2 (New Series), Springer-Verlag, Berlin.

PITMAN, M. G. (1977) Ion transport into the xylem, *Ann. Rev. Plant Physiol.*, **28**, 71–88.

SLATYER, R. O. (1967) *Plant-Water Relationships*. Academic Press, London & New York.

Other references

BAKER, D. A. and WEATHERLEY, P. E. (1969) Water and solute transport by exuding root systems of *Ricinus communis*, *J. exp. Bot.*, **20**, 485–95.

BELL, C. W. and BIDDULPH, O. (1963) Translocation of calcium. Exchange versus mass flow, *Plant Physiol.*, **38**, 610–614.

CLARKSON, D. T. and SANDERSON, J. (1971) Relationship between the anatomy of cereal roots and the absorption of nutrients and water, *A. R. C. Letcombe Lab. Ann. Rep.*, 1970, 16–25.

FERGUSON, I. B. and BOLLARD, E. G. (1976) The movement of calcium in woody stems, *Ann. Bot.*, **40**, 1057–1065.

HARRISON-MURRAY, R. S. and CLARKSON, D. T. (1973) Relationships between structural development and the absorption of ions by the root system of *Cucurbita pepo*, *Planta (Berl.)*, **114**, 1–16.

HEINE, R. W. (1970) Absorption of phosphate and potassium ions in poplar stems, *J. exp. Bot.*, **21**, 497–503.

HELLKVIST, J., RICHARDS, G. P. and JARVIS, P. G. (1974) Vertical gradients of water potential and tissue water relations in sitka spruce trees measured with the pressure chamber, *J. appl. Ecol.*, **11**, 637–668.

KLEPPER, B. and KAUFMANN, M. R. (1966) Removal of salt from xylem sap by leaves and stems of guttating plants, *Plant Physiol.*, **41**, 1743–1747.

MARSHALL, C. and WARDLAW, I. F. (1973) A comparative study of the distribution and speed of movement of ^{14}C assimilates and foliar applied ^{32}P-labelled phosphate in wheat, *Aust. J. biol. Sci.*, **23**, 1–13.

MILTHORPE, F. L. and MOORBY, J. (1979) *An Introduction to Crop Physiology*, 2nd ed., Cambridge University Press, Cambridge.

MORRISON, T. M. and HEINE, R. W. (1966) Transport of minerals in tree xylem, *Ann. Bot.*, **30**, 807–819.

ROBARDS, A. W. and CLARKSON, D. T. (1976) The role of plasmodesmata in the transport of water and nutrients across roots, pp 181–199, in Gunning, B. E. S. and Robards, A. W. (eds), *Intercellular Communication* in *Plants: Studies on Plasmodesmata*. Springer-Verlag, Berlin.

Chapter 4

Transport systems and plant growth

4.1 Introduction

In the first three chapters I have been concerned with providing
descriptions of the tissues involved in transport and the way in which
sugars, water and ions move through them. I now want to provide a
brief account of how these transport systems react to the influence
of various environmental and internal factors and the effect of
these interactions on growth.

The growth of a plant depends on the supply of metabolites to
meristems. It is not often, however, that growth is controlled by a
single factor. Rather, growth is the final product of the complex
interrelationships between the processes in the meristems and the
effects of internal and external factors on these processes and the
supplies of metabolites. For example, the light received by the crop (or
lack of it) is a major factor governing the growth of glasshouse crops in
the winter months in the UK. Nevertheless, in a period when the
maximum flux density of photosynthetically active radiation is often
only 40 Wm^{-2}, the rate of growth can be increased by changes in the
CO_2 concentration, the temperature and the supply of mineral
nutrients.

Because of these complex interrelationships, it is, perhaps, not
surprising that there are few well authenticated instances of
deficiencies in the transport systems restricting plant growth. On the
contrary, there are many examples which show that the systems are
sufficiently flexible to respond to short-term fluctuations in both the
amounts to be moved and the distribution patterns, and to long-term

extra demands placed upon them during evolution. The final section of this chapter will, therefore, give an account of how the transport systems react to the changing demands made upon them during the growth and development of higher plants.

4.2 The effect of environmental and other factors on transport systems

Environmental factors can influence the transport processes in a wide variety of direct and indirect ways. It is difficult, therefore, to give a simple but comprehensive account of the effects of the different factors. Instead, I will briefly discuss the effects of each factor in turn. Much of this is obvious, even trite, but deserves stating. Some of the less obvious influences and interactions will be described in greater detail in the subsequent sections where particular physiological situations are considered.

4.2.1 Water
A decrease in the supply of water and consequent reduction in the leaf water potential can lead to smaller amounts of assimilates entering the phloem. Part of this effect, and possibly most, can be attributed to a reduced rate of photosynthesis but there can also be effects on the movement through the phloem although these appear to vary with the species.

For example, in one series of experiments there was a 20 per cent decrease in the amount of carbon assimilated by droughted potato plants but there was no reduction in the proportion of the assimilated carbon translocated from the leaves (Table 4.1). The reduction in the amount of carbon being converted to polysaccharides is typical of

Table 4.1 The effect of drought on photosynthesis and distribution of assimilated carbon in potato plants bearing tubers. The data are expressed relative to a net carbon fixation of 100 arbitrary units by the control plants. The data in parenthesis are the values for droughted plants expressed as a percentage of the carbon fixed by these plants (after R. Munns and C. J. Pearson, 1974).

	Control plants	Droughted plants
Net photosynthesis	100	80
Respiration	6	5 (6.3)
Labelled carbon retained in leaf exposed to $^{14}CO_2$		
Ethanol-soluble	16	21 (26.3)
Polysaccharides	15	4 (5.0)
Translocated	63	50 (62.5)

Table 4.2 The distribution of ^{14}C-labelled assimilate in control and water stressed wheat plants 24 h after exposure of the flag leaf to $^{14}CO_2$. The values are the means of eight replicates (± standard errors). (After I. F. Wardlaw, 1967).

Plant part	Control plants	Stressed plants
Flag leaf	26.4 ± 3.8	57.4 ± 4.3
Ear	34.7 ± 3.9	33.7 ± 3.5
Top internode	5.2 ± 0.9	3.0 ± 0.9
Lower internodes	17.5 ± 2.1	2.9 ± 1.2
Roots	16.3 ± 2.7	3.1 ± 0.6

plants suffering from a water stress. In contrast, there was a marked effect of drought on translocation in wheat, the amount of carbon exported from the flag leaf being reduced by about 40 per cent. The amount of assimilated carbon which was transported to the dominant sink, the ear, remained about the same, but there was a marked reduction in the amount of carbon translocated to the basal regions of the plant. (Table 4.2)

The effect of a water deficit on the amount of assimilates moved by a plant could be exerted because of changes in both the stomatal and residual resistances to CO_2 uptake. The former are the most common, but effects on the residual resistance can be seen in some species. The effects on processes in the phloem are less well understood. However, the greater osmotic concentration in the sieve cells than in the surrounding tissues would suggest that the transport processes in the phloem might be sensitive to changes in the water potential gradients since the latter could influence the exchange of water between these tissues and would certainly affect any transport resulting from the operation of a pressure flow mechanism. If the latter operated, the movement of water through the phloem could be considerable, but this is often overlooked because of the usually greater movement through the xylem (Fig. 4.26).

The effects of plant water deficits on the movement of water through the xylem are readily explicable in terms of the ways in which the gradients in water potentials are established and maintained. However, these effects also have repercussions on the movement of ions which can be further complicated by consequent changes in the growth substances in the plant. For example, abscisic acid (AbA) production by the roots, and supply to the leaves, is increased when the plant experiences a water deficit and AbA appears to stimulate the transport of ions from the root to the shoot whilst having little effect on ion uptake.

Fig. 4.1 Diurnal fluctuations in the insolation and the uptake of nitrogen, potassium and ▶ water by a crop of tomatoes growing in a glasshouse in a flowing nutrient solution (unpublished data of P. J. Adams).

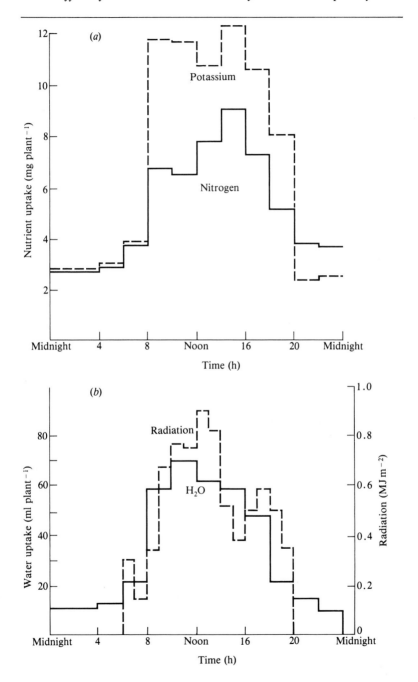

There is also evidence of further interactions in the leaves. The guard cells of the stomata open because of an increase in the osmotic potential which seems to result from an influx of potassium ions from the surrounding cells. Stomatal closure is accompanied by an efflux of K^+ ions and can be induced by the application of AbA. There is, therefore, an apparently close interdependence between plant water relations, ion transport and growth substance effects but the nature of these relations is still not clear.

4.2.2 Light

The rate of ion uptake by a plant is a function of the flux of short wave radiation to which the plant is exposed (Fig. 4.1). This can be related in part to the effect of radiation on the rate of transpiration. In addition, however, in these experiments, there was continued uptake of nitrogen and potassium in periods of darkness and this was probably dependent on a supply of energy from respiration and, hence, the transport of assimilates from the aerial parts of the plant. A similar pattern was shown in experiments on nitrate uptake (Fig. 4.2) where again it was concluded that the most important factor controlling the rate of uptake was the supply of assimilates from the leaves. The subsequent transport of ions and water through the xylem is influenced by the insolation because of the sensitivity of the rate of transpiration to the flux density of radiation.

The effect of light on the rate of photosynthesis leads to obvious effects on the amount of assimilates moving through the phloem. In addition, however, there are more subtle effects of light on the movement of assimilates into the sieve tubes. Light can decrease the recycling of starch and mobile carbon in the leaf, as indicated by the slower rate of export of carbon from the leaves of *Amaranthus* after about one hour's transport (Fig. 4.3.). However, this effect is not specific, and can also be induced by other treatments which inhibit photosynthesis. (See section 2.3.3 for the interpretation of this type of data.) This analysis is not unequivocal. It assumes uniform mixing of the labelled carbon in the various compartments in the leaf and this may not be true in the immobile compartment where much of the carbon is probably in the form of starch. The starch seems to be laid down in, and removed from, the starch grains in concentric layers. Because of this, some observations suggest that in some species the carbon assimilated late in a photoperiod appears to be transported from the leaf earlier in the following dark period than does carbon which has been assimilated at the beginning of the light period. That is, there appears to be a 'first in, last out' situation.

Light has a major effect on the speed of translocation. The speed, as measured in experiments using ^{11}C-labelled assimilates, increased rapidly as the plant entered a photoperiod and attained a maximum

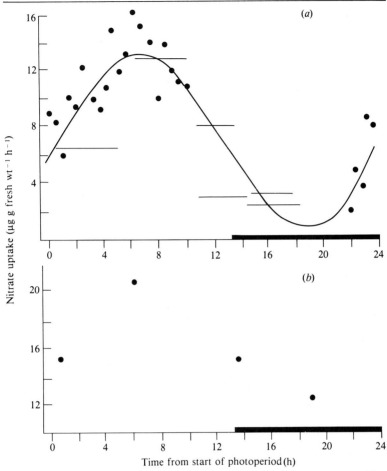

Fig. 4.2 Diurnal fluctuations in the rate of nitrate uptake by *Capsicum annuum*. The data in (*a*) were from 2 experiments (● , —) together with a fitted harmonic line. Each point is the mean of two 4 plant samples. (*b*) Shows the nitrate uptake in another experiment determined using ¹⁵N. Each point is the mean from 2 plants. The heavy lines on the abscissae indicate the dark periods (after C. J. Pearson and B. T. Steer, 1977).

value after only 2–3 h in the light. It decreased progressively throughout a subsequent dark period (Fig. 4.4). These effects appear to be related to an increase in the sucrose concentration in the leaf during the photoperiods and its decline in the dark. It has been suggested that when the concentration of sucrose in the bulk leaf tissue of maize plants is less than about 7 per cent a normal concentration-sensitive loading process operates. Above this concentration the

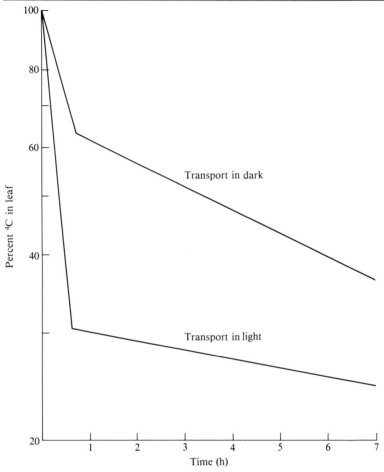

Fig. 4.3 The change with time in the amounts of ^{14}C remaining in the leaves of *Armaranthus* plants which had been exposed to $^{14}CO_2$ for 2 minutes and either maintained in the light or transferred to the dark (unpublished data of J. Moorby).

sucrose gradients in the leaf might influence loading directly thus causing a biphasic response of speed to sucrose concentration.

In addition to these effects of light on the amount of assimilates moving and the speed, light can also affect the distribution from leaves of both assimilated carbon and ions. For example, both pretreatment with light before exposure to $^{14}CO_2$, and during the exposure, increased the proportion of the ^{14}C which moved downwards to the old leaves, tillers and roots of *Lolium* and *Sorghum* (Table 4.3). In the latter

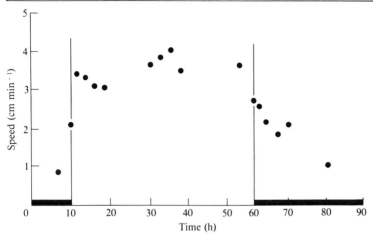

Fig. 4.4 The speed of ^{11}C translocation in maize in a sequence of 12 h in the dark, 50 h in the light and a further 20 h in the dark. The dark periods are indicated by the heavy lines on the abscissa (after J. Moorby, J. H. Troughton and B. G. Currie, 1974).

instance pretreatment in the light was more effective at increasing movement downwards than was exposure to light during the exposure. Similarly, the amount of foliar applied ^{137}Cs exported from *Pisum* leaflets was increased by exposure to light, and again, there was an increase in the proportion moving down the plant (Table 4.4).

Table 4.3 The effects of light before and during the experiment on the distribution of [14]CO_2-labelled assimilates in *Lolium temulentum* and *Sorghum sudanense* (recalculated from I. F. Wardlaw, 1976).

The pretreatment was growth for 3 days at low irradiances, 20 Wm^{-2} for 8 h per day (400-720 μm), and high irradiances, 96 Wm^{-2} for 8 h per day. Each value is the mean of six observations ± standard errors and is the percentage of the ^{14}C exported from the treated leaf in 4 h.

	Pretreatment	Movement	Light during dark	[14]C distribution 108 Wm^{-2}
Lolium	20 Wm^{-2}	Upwards	65 ± 3.4	48 ± 2.7
		Downwards	35 ± 4.9	52 ± 4.1
	96 Wm^{-2}	Upwards	33 ± 2.3	24 ± 3.8
		Downwards	67 ± 3.4	76 ± 4.5
Sorghum	20 Wm^{-2}	Upwards	62 ± 1.7	58 ± 2.6
		Downwards	38 ± 2.0	42 ± 5.2
	96 Wm^{-2}	Upwards	37 ± 2.7	33 ± 2.9
		Downwards	63 ± 5.6	67 ± 6.5

115

Table 4.4 The movement of ^{137}Cs from *Pisum sativum* leaves in response to light and darkness. The amount of ^{137}Cs exported from the treated leaflet is expressed as a percentage of that absorbed and the amounts in the different parts of the plant as percentages of that which moved from the treated leaflet (after J. Moorby, 1964).

	% [137]Cs exported	Rest of treated leaf	Apical parts of plant	Basal parts of plant
Plant in light	29.5	7.4	34.0	58.6
Plant in darkness	6.8	28.2	38.8	33.0
Least significant difference P = 0.05	2.8	4.9	14.1	15.3

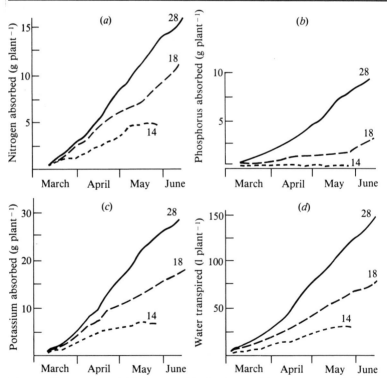

Fig. 4.5 The amounts of (*a*) nitrogen, (*b*) phosphorus, (*c*) potassium and (*d*) water absorbed over a period of 4 months by tomatoes growing in flowing nutrient solutions maintained at 28 °C (———), 18 °C (— — —) and 14 °C (- - - -) (unpublished data of J. Moorby).

4.2.3 Temperature

The effects of temperature on transport processes can be considered in terms of both long and short-term effects on movement through the xylem and phloem and effects which influence either sources, sinks, or the regions between the two.

Long-term effects are usually exerted because of changes in the patterns of growth within the plant. For example, when plants are grown at low root temperatures they produce smaller root systems than do plants where the root temperature is greater. These differences can result in major changes in the amounts of water and ions absorbed by the plants (Fig. 4.5).

In addition to these long-term effects, there are short-term effects, of low root temperatures which become apparent very soon after imposition of the treatment. Low root temperatures appear to reduce the stomatal resistance and leaf water potential by decreasing the permeability of the roots to water and, in turn, the supply of water to the leaves. There are also short-term effects on ion uptake which can lead to complex interactions if the root temperature is subsequently changed (Fig. 4.6). These results can be attributed in part to effects of

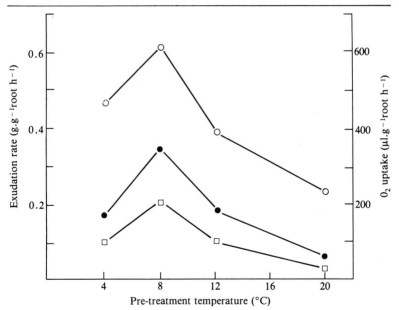

Fig. 4.6 The effect of pretreatment of whole plants for 72 h at a range of root temperatures on the rates of xylem exudation and oxygen uptake of excised barley roots at 20 °C. Exudation rate during first hour of experiment (●——●); mean exudation rate over 24 h (□——□); oxygen uptake (○——○) (after D. T. Clarkson, M. G. T. Shone and A. V. Wood, 1974).

117

the treatments on the rate of respiration (cf. the rate of oxygen uptake in Fig. 4.6) but other, less well understood effects on growth substances in roots might also be involved.

The long-term effects of temperature on the rates of growth of sinks must necessarily have associated with them changes in the patterns of assimilate movement. For example, when tomato fruit are warmed the amount of assimilate moving from nearby leaves is increased, as is the rate of turnover of carbon in the leaf (Fig. 4.7). There is also an increase in the speed at which assimilates move into the fruit. In

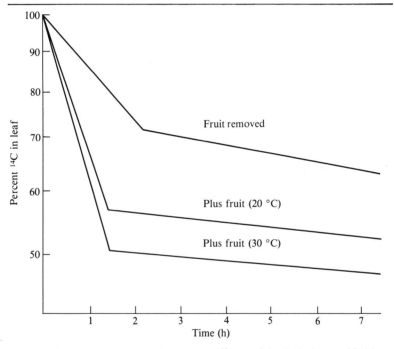

Fig. 4.7 The change with time in the amount of ^{14}C remaining in the leaves of fruiting tomato plants which had been exposed to $^{14}CO_2$ for two minutes when the fruit was maintained at 20 or 30 °C or removed (after J. Moorby and P. D. Jarman, 1975).

contrast, when the fruit is cooled, it may, at certain stages of its growth, change from being a net-importer of carbon to a net exporter (Table 4.5). This change is associated with an increased accumulation of sucrose in the fruit but it is not known whether this reflects a decrease in the rate of utilization of imported sucrose or an increased synthesis of sucrose from other materials.

A more transient effect is seen when part of the transport system

Table 4.5 The effects of temperature on the rates of respiration and movement of carbon into (+) and out of (−) tomato fruit and the concomitant changes in total carbon (△ carbon) and sucrose (△ sucrose) over a period of 48 h. (After A. J. Walker and L. C. Ho, 1976).

Fruit Temperature (°C)	Initial carbon Content (g)	△ Carbon (mg 48 h^{-1})	Carbon respired (mg 48 h^{-1})	Carbon translocated (mg 48 h^{-1})	△ Sucrose (mg 48 h^{-1})
5	1.0 ± 1.3	+ 96.1 ± 15.8	4.5 ± 0.5	+100.6 ± 15.4	+28.3 ± 7.2
25	1.0 ± 1.4	+234.3 ± 10.6	19.1 ± 2.4	+253.4 ± 13.0	− 3.4 ± 7.7
5	1.6 ± 1.9	− 84.5 ± 28.8	5.6 ± 0.5	− 78.9 ± 28.8	+36.8 ± 8.6
25	2.0 ± 2.2	+126.3 ± 16.3	34.5 ± 13.0	+160.8 ± 3.8	− 4.8 ± 4.3

between the source and sink is cooled (Fig. 2.19). In addition to the short-term effect on speed there is also a reduction in the rate of translocation. However, both parameters recover a short time after the low temperature treatment is imposed and regain their original values even if a large part of the pathway is treated (Fig. 4.8).

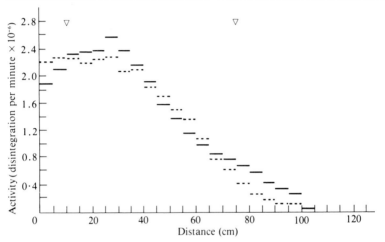

Fig. 4.8 The distribution patterns of ^{14}C-labelled assimilates in willow stems maintained at 23 °C (——) or cooled to 0 °C (- - - -) over a 65 cm length indicated by the arrows. The data are the means of 7 experiments (after B. T. Watson, 1975).

4.2.4 Growth substances

It is difficult to critically evaluate the effects of growth substances on transport, or any other plant process, because of the lack of adequate quantitative information on the concentrations of these substances in the various tissues, their rates of production, turnover and transport, and exactly how they bring about the responses they undoubtedly cause.

For example, IAA and related auxins have profound effects on the growth of various organs and can influence cell expansion and patterns of branching in roots and shoots. The generally accepted explanation is that the auxin is produced in the meristems and moves from these in a polar fasion, i.e. predominantly away from the meristem, through tissues near the cambial zone, e.g. differentiating secondary xylem elements. This polar movement has been invoked to explain many correlative growth phenomena such as apical dominance and the onset of secondary xylem production in woody plants. What is often overlooked, however, is that analyses of phloem sap show that the concentration of IAA is often at physiological levels and that this material is free to move throughout the vascular system of the plant. Similarly, gibberellins, kinins and abscisic acid have all been detected in both the phloem and xylem saps. In these circumstances it is difficult to be sure of the extent to which the response to an applied growth substance is localized within the treated region, and, indeed, what are the physiologically active concentrations in the 'target' tissues.

It is not surprising, therefore, that many of the responses of the transport systems to applied growth substances tend to be relatively unspecific. In many instances they can be explained most easily in terms of the applied materials initiating a sink into which transport occurs. For example, the results in Table 4.6, show the apparently

Table 4.6 The effect of IAA, kinetin and gibberellic acid (GA) on the movement of ^{32}P into decapitated second internodes of *Phaseolus vulgaris*. The concentration of all growth substances was 1000 ppm in anhydrous lanolin. The ^{32}P was applied to a flap of epidermal tissue on the first internode (after A. K. Seth and P. F. Wareing, 1967).

Treatment	Mean count rate (counts min^{-1})	Standard error $(+)$
Control	23.6	4.7
GA	84.8	16.9
Kinetin	58.1	17.8
IAA	453.4*	143.1
IAA + GA	849.6*	278.6
IAA + Kinetin	889.4*	255.4
IAA + Kinetin + GA	1943.1*	312.4

* Significantly different from control $P > 0.01$

additive effect of IAA, kinetin and gibberellic acid even though the errors are large. More recent experiments, however, again demonstrated the transport of ^{14}C-labelled assimilates into regions treated with IAA but could detect no effect of the IAA treatment on the rates of sucrose consumption, respiration and protein synthesis, ie. there was no apparent increase in sink activity. Furthermore, there was some evidence that the effect of IAA could be detected at positions remote from the point of application and that it enhanced the movement of material out of the phloem. The effect of growth substances on transport through the phloem therefore remains obscure and requires much more investigation.

4.3 Transport systems and growth

It is obvious that the rate of growth of a meristem is a function of the amounts of organic and inorganic nutrients available. Because the

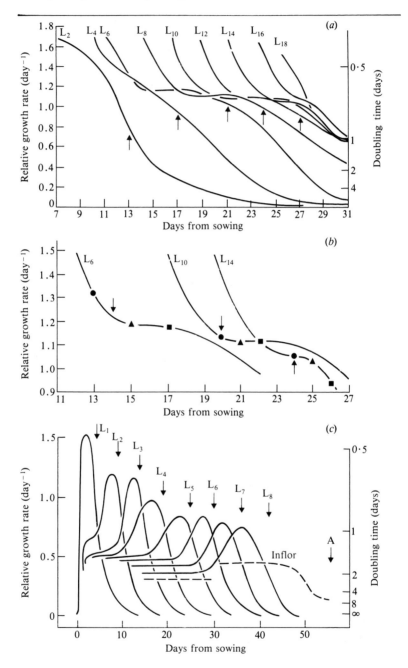

nutrient requirements of actively growing meristems are large relative to their size, the opportunity for them to store significant amounts of material is small. Hence, their growth is dependent on the continual supply of nutrients from other parts of the plant. A discussion of the nature of these supplies, and how they are controlled, forms the basis of the remainder of this chapter. In order to do this we have to examine the developmental processes which operate in the various meristems and how they are co-ordinated. The discussion will then be developed to examine transport into storage organs and fruit.

4.3.1 Transport and the growth of stem apices and leaves

The first vascular tissue to differentiate in a stem and in leaf primordia is protophloem and this is only detectable several internodes behind the stem apex. Xylem does not appear until several more plastochrons have elapsed. The apex and leaf primordia are dependent, therefore, initially on nutrients which reach them by diffusion through the meristematic tissues after transport through the phloem. The tissues supplied in this way may be be bigger than is at first apparent, since it is not clear at what point in the differentiation of phloem it becomes capable of supporting translocation.

Diffusion is a very inefficient means of transporting nutrients through a mass of rapidly growing cells and there sometimes appears to be some limitation of the rates of growth of small leaf primordia. For example, in wheat, there is a marked increase in the relative growth rates of primordia, and in tobacco they are maintained at a uniform rate, after the differentiation of the first vascular traces connecting the primordia to the vascular tissue of the stem (Fig. 4.9).

The ways in which transport processes are involved in growth are dependent on the mechanics of growth in the organ in question. Major differences are seen between the growth of leaves of monocotyledonous and dicotyledonous plants. The tissues of the former are all produced from a basal meristem. By the time the new tissue at the tip of the leaf has emerged from the enclosing sheath of the previous leaf it is fully expanded and mature, even though the more basal parts are still being formed. The most active phase of cell division in the leaves of

---

◀ **Fig. 4.9**(*a*) The relative growth rates and doubling times of tobacco leaf primordia. The arrows indicate the times of emergence of the primordia from the apical bud. The data from alternate leaves are omitted.

(*b*) The relative growth rates of tobacco primordia. The arrows show the time of first appearance of the leaf lamina and the times of appearance of the first phloem (●), and the start (▲) and final (■) differentiation of sieve tubes (after R. V. Hannam, 1968).

(*c*) The relative growth rates and doubling times of wheat leaf primordia. The arrows indicate the times of emergence of the leaves and A the time of anthesis. The first protophloem elements in each primordium were detected about the time of rapid increase in the relative growth rate (after R. F. Williams and C. N. Williams, 1968).

dicotyledonous plants takes place in the apical bud. As the leaf emerges from the bud it starts to increase in area. At the same time the rate of cell division starts to decline, and the increase in area results mainly from cell expansion. Within this overall framework there is a gradation in the rate of differentiation, tissues in the distal parts tending to mature before those in the proximal regions.

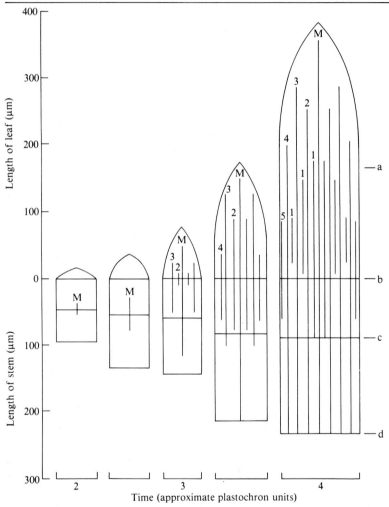

Fig. 4.10 The changes with time (plastochron units after initiation) in the mean lengths of the primordia of wheat leaves (a − b), of the stem in the node of insertion (b − c), and to the base of the second leaf below (c − d) and of each vascular bundle present. M is the median bundle and the numbers the lateral bundles (after J. W. Patrick, 1972a).

These two types of differentiation lead to different patterns of development of the transport systems. The phloem which is formed in the leaf primordia of monocotyledonous plants does so in isolation from existing phloem in the stem and differentiation proceeds both acropetally and basipetally. The median (i.e. central) bundle is the first to appear followed by the lateral bundles, (Fig. 4.10) and the downward differentiation soon extends to the second node below that of insertion of the leaf. It is only at this second older node that a complete nodal plexus is formed and it is apparently for this reason that most of the assimilate imported by a developing leaf comes from the second leaf below. The completion of this vascular pathway usually coincides with the maximum rates of photosynthesis of the older leaf and growth rates of the younger (Fig. 4.11).

The continued division in the basal meristem of a developing leaf presents problems in maintenance of continuity of the phloem and sometimes breaks occur. Further, there are also problems in the export of assimilates. When the apical parts of a leaf of a monocotyledenous plant emerge from the leaf sheath they are mature and capable of photosynthesis. After the immediate needs of these parts have been satisfied some of the surplus is used to form starch and the remainder has to be exported. There is, therefore, a requirement for functional phloem through which this can take place. Tracer experiments have shown the existence of such a transport system with ^{14}C labelled assimilates moving from the apical regions of a half-expanded wheat leaf to other parts of the plant. Simultaneously, assimilates are transported from older leaves to the meristematic regions of the growing leaf and these can account for about 80 per cent of the final dry weight of the leaf lamina. It follows that the remaining 20 per cent is supplied by photosynthesis in the leaf itself. In contrast, about 50 per cent of the final dry weight of a leaf sheath can be attributed to the transport of assimilate from the attached lamina.

The control of the transport processes in the basal meristem must be complex. It will be seen from discussions in this chapter that as a general rule sinks are supplied mainly from the nearest available sources with adequate vascular connections. In the leaves of monocotyledonous plants, however, most of the material which is transported basipetally through the sieve tubes passes directly through the meristem to other, more distant, growing points. This occurs at the same time that assimilates moving acropetally from older leaves are being unloaded into the meristem to support the growth of the leaf. That is, the control of the unloading processes is dependent on the direction of translocation through the sieve tubes. How this is accomplished is not known, but it could be associated with transport in the two directions in separate vascular bundles.

Examination of quantitative aspects of the export of assimilates

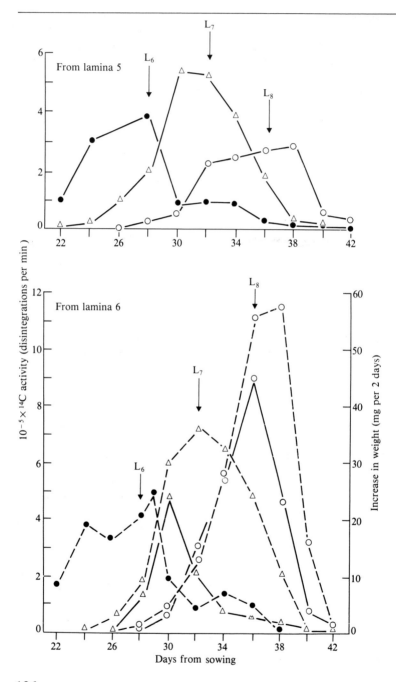

reveal further uncertainties. From estimates of the rates of photosynthesis and dry weight accumulation in different parts of the leaf it is possible to calculate the rate of translocation from these regions. This decreases towards the base of the leaf unlike the amount of dry matter moving through the leaf which increases (Fig. 4.12*a* and *b*). When these estimates are coupled with measurements of the cross-sectional area of the phloem in the major and minor vascular bundles along the length of a leaf (Fig. 4.12*c* and *d*) it is possible to estimate the rate of translocation per unit area of phloem. This also increases markedly (Fig. 4.12*e* and *f*). These latter estimates assume that translocation takes place through either all the phloem or only the main veins. Translocation through the main veins may be the more realistic since there is some evidence that the major role of the minor and transverse veins is in loading the major veins and the storage of assimilates. The data also show that the greater rate of photosynthesis in the C_4 grass is associated with a larger rate of translocation.

If, as seems likely, translocation involves a flow of solution these observations raise difficulties of interpretation in relation to the mechanism of translocation. Since rate $(g\,m^{-2}\,s^{-1})$ = speed $(m\,s^{-1})\times$ concentration $(g\,m^{-3})$ the increasing rate along a leaf lamina suggests that either the speed of translocation, or the concentration of the solution, or both, should also increase. Estimates of speed along the leaf lamina show no evidence of any change. Further, any increase in concentration would tend to favour movement towards the tip of the leaf.

Events in the leaves of dicotyledonous plants are somewhat different from those described above. All the growth which takes place whilst the primordia are within the apical bud must be dependent on the transport of assimilates from the seed or older expanded leaves. It is not surprising, therefore, that growth during this time and, in particular, the number of cells produced, is a positive function of the irradiance to which the older leaves are exposed. Cell division does not stop when the leaf emerges from the apical bud and it, and the net import of assimilates, continue until the leaf is about a half to two-thirds of its final area. Before this time the leaf starts to export assimilates. Export usually begins when the leaf is about a quarter of its final area, but in tomato it starts when the leaves are only one-twentieth of their final area.

These overall changes are the result of more detailed changes in leaf development which show some resemblance to events in the leaves of

---

◄ **Fig. 4.11** The amount of ^{14}C-labelled assimilates transported at various times from leaves 5 and 6 of wheat to leaves 6 (●——●), 7 (△——△), and 8 (○——○) and the rates of growth of these leaves (same symbols, dotted lines). The arrows indicate the times of complete emergence of the laminae (after J. W. Patrick, 1972*b*).

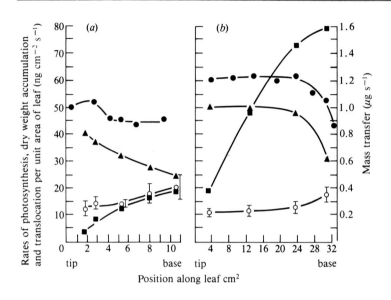

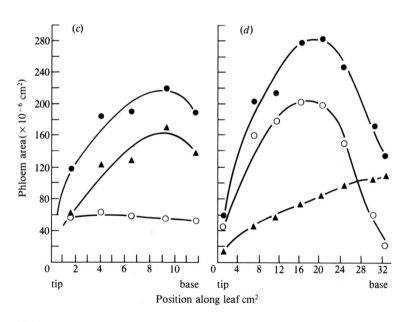

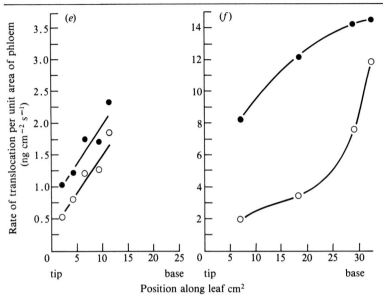

◀ **Fig. 4.12** Various parameters of the transport systems of the C₃ grass *Lolium temulentum* (*a*,*c* and *e*) and the C₄ grass *Panicum maximum* (*b*,*d* and *f*). (*a*) and (*b*) show the rates of photosynthesis (●——●), dry weight accumulation (○——○), and translocation by difference (▲——▲) from unit area of leaf at different positions along the leaf and the rate of mass transfer (■——■). (*c*) and (*d*) show the area of the total phloem (●——●), the major veins (▲——▲) and the minor veins (○——○) and (*e*) and (*f*) the rate of translocation assuming transport through all the phloem (○——○) or only the major veins (●——●) (after W. M. Lush, 1976).

monocotyledonous plants. As in the latter, the differentiation and maturation of the tissues is completed first in the most distal parts of the leaf and the changes spread down to the more basal regions. The main veins of the leaf differentiate acropetally, and it has been suggested that these are required to support the import of assimilates from older leaves. In contrast, the differentiation of the phloem in the minor veins takes place basipetally, and at a later stage of leaf growth. In species with bicollateral bundles the adaxial phloem appears to mature before the abaxial.

The changes in the transport of assimilates associated with these events are what might be expected. The region where there is the most active accumulation of imported assimilates is initially in the distal regions. As the photosynthetic abilities of the tissues increase, the demand for imported assimilates declines and the apical parts start to export materials. This change from the import to the export of assimilates gradually progresses to the more basal regions of the leaf.

The onset of export appears to coincide with the development of mature sieve tubes in the minor veins which, as in monocotyledonous plants appear to be concerned with the loading of assimilates into the phloem. This general sequence of events has been shown to take place in more than one species. There is, however, doubt about whether the distal parts of a leaf supply assimilates to the younger more basal parts. This uncertainty could have arisen because of true interspecific differences but could equally result from sampling difficulties in tissues which are changing rapidly.

The major factor controlling the pathways of transport between leaves seems to be the existence of functional vascular connections. Hence, the movement of assimilates into and out of a leaf can be related to the phyllotaxy of the plant. The major connections between the leaf traces coming from a petiole and the vascular bundles in a stem tend to be at the nodal plexus one node below the level of leaf insertion. Hence, most of the assimilates which eventually move up the plant from a leaf initially move down to the node below that of insertion of the leaf. Detailed anatomical and physiological experiments have shown that in some species there is an even closer level of control than that described, with assimilates formed in a particular region of a leaf moving preferentially to a specific part of another.

The preceding discussion has been concerned with the interrelations between leaf growth and the translocation of assimilates; the interrelations with, and consequences of the movement of mineral nutrients are rather different. Leaf expansion is particularly sensitive to restrictions in the supply of mineral nutrients, possibly because of a requirement for ions to act as osmotica in cell expansion. In addition, probably because of restrictions in the supply of ions during the early growth of primordia before good vascular connections are established, nutrient deficiencies can lead to decreases in the number of cells per leaf.

A consequence of the role of ions in leaf expansion is the later onset of the export of ions than carbohydrates. The former usually only starting when leaf expansion is completed (Fig. 4.13). This export takes place through the phloem, as does the initial import. As the leaf expands, and transpiration begins, an increasingly greater proportion of the ions enter the leaf through the xylem and this continues throughout the life of the leaf. The amount of ions in a leaf at any one time is the difference between those imported and exported and tends to increase through the period of leaf growth and for some time after growth stops. Eventually, most leaves become net exporters of ions such as phosphate and potassium which can move through the phloem and the amounts of these ions in the leaf start to decline. The decrease is usually associated with other phenomena related to leaf senescence

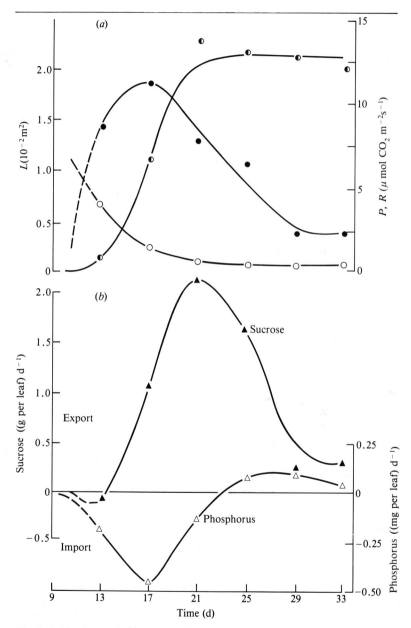

Fig. 4.13 The changes in (a) Leaf area (◑——◑) and rates of photosynthesis (●——●) and respiration (○——○) and (b) the rates of import and export of sucrose (▲——▲) and phosphorus (△——△) of the second leaf of cucumber (after J. M. Hopkinson, 1964).

such as a decrease in the chlorophyll content. In contrast to these events, the amount of calcium in leaves usually increases until abscission and the amounts of the slightly mobile ions, such as magnesium, tend to remain approximately constant.

The ions which are exported move to younger growing tissues and the export, which is initiated and controlled in the leaf, seems to be a form of prudent housekeeping by the plant and does not appear to be a response to a demand from other organs. The exact timing of the onset of net export by a leaf is dependent on many factors and occurs earlier in plants with a restricted supply of nutrients than in those with an adequate supply. In experiments with barley, for example, the rate of export of phosphorus from leaves was initially independent of the rate of supply of phosphorus to the roots, but eventually declined as the effects of phosphorus deficiency became more severe. In plants grown with an adequate supply of phosphorus the rate of export remained constant over a long period and the amounts of phosphorus in the exporting leaves showed a net increase because of the continued import.

The concentrations of mineral nutrients in leaves are dependent not only on the amounts of ions moving in and out but also the age of the leaf and the general carbon economy. The concentrations are usually high in primordia and then fall to approximately constant values as the leaves expand. For example, the concentrations of nitrogen and phosphorus in the primordia of wheat leaves are usually about 100 and 15 mg g^{-1} dry weight, but fall to about 30 and 4 mg g^{-1} dry weight respectively in mature leaves. They can finally decrease even further during leaf senescence.

At any one time the actual concentrations are determined by the processes described above. In stable situations events are sufficiently predictable for leaf analysis to be a useful technique for diagnosing the nutrient status of plants. In less stable situations, however, individual measurements can be misleading. For example, under conditions of nutrient deficiency, the concentrations of phloem-mobile ions are maintained in the younger tissues at the expense of the early senescence of the older. In contrast, the concentrations of calcium and sulphate can be very low in young tissues and at normal levels in the older leaves which are usually analysed.

The transport of nitrogenous compounds is rather more complex then is that of the other mineral nutrients. The requirement in the plant for reduced nitrogen, coupled with the absorption of large amounts of nitrogen as nitrate, necessitates the reduction of the latter. This is accomplished by the enzyme nitrate reductase which in some species is restricted to specific regions of the plant.

For example, in *Pisum*, nitrate reductase is found in the roots and is absent from the leaves. Absorbed nitrate is, therefore, reduced in the roots to compounds such as asparagine and glutamine which are used there or transported to the leaves and other growing organs through the xylem. Free nitrate ions can be found in the aerial parts of this type of plant only when they are supplied in such large amounts that the

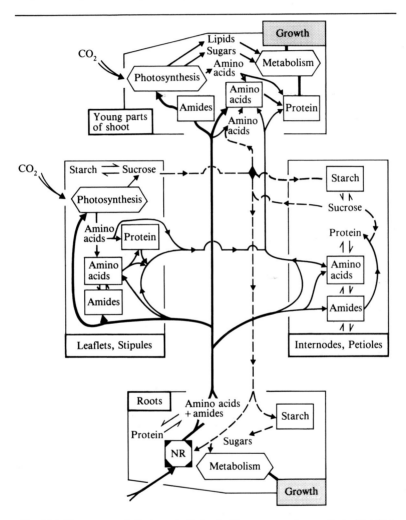

Fig. 4.14 The pattern of nitrate assimilation in *Pisum arvense* L. supplied with a low level of nitrate to ensure nitrate reductase [NR] was restricted to the roots. Under these conditions there is no storage of free nitrate. The movement of the reduced nitrogen through the xylem is indicated by the heavy solid line. The movement through the phloem of assimilates and reduced nitrogen is indicated by thinner broken and solid lines respectively (after W. Wallace and J. S. Pate, 1967).

enzyme cannot reduce all the ions absorbed. The reduction in the roots is dependent on a continual supply of assimilates from the leaves to provide the necessary energy and the carbon skeletons for the amines etc. (Fig. 4.14).

In contrast, in plants such as *Xanthium*, the nitrate reductase is located in the leaves. In such species the absorbed nitrate moves to the leaves through the xylem before being reduced and exported to the remainder of the plant through the phloem. The roots are unusual in being unable to use the nitrates which they absorb from the external medium. Instead they have to rely on reduced forms of nitrogen supplied from the leaves in the phloem (Fig. 4.15).

Another factor which can change the uptake and transport of nitrogen is the ability of the plant to fix atmospheric nitrogen, either directly as in some tropical grasses or by a symbiotic partnership with microorganisms such as *Rhizobium* in root nodules. It is possible to collect samples of both xylem and phloem sap from several positions on nodulated plants of *Lupinus albus*. These can be analysed and, together the results of gas exchange measurements, used to construct a balance sheet of the uptake and utilization of carbon and nitrogen (Fig. 4.16). This shows several surprising features and emphasizes the extent to which the xylem and phloem operate together as an integrated transport system which gradually changes as the plants age (Table 4.7). For example, the vegetative apices and fruit received 75 per cent and 40 per cent of their nitrogen and about 15 per cent and 6 per cent of their carbon respectively via the xylem. A third of the carbon translocated to the nodules in the phloem is returned to the shoot as products of nitrogen fixation. Almost half of the nitrogen incorporated into the growing nodules was transported from the shoot in the phloem. Much of the nitrogen which moved into the fruit in the phloem was transferred to that tissue from the xylem in the stem and more than half of the nitrogen which moved into the leaves in the xylem was later exported via the phloem.

Nitrogen not immediately required for growth appears to enter a storage pool which can be drawn on to maintain growth if uptake is reduced at a later time. This can be seen in the results of some experiments with *Lolium perenne* grown in a flowing nutrient solution where the amount of nitrate absorbed could be monitored and controlled precisely. Here, when the supply of nitrate was stopped for nine days, so making the plants dependent on stored nitrate, there was little effect on plant growth but there was a marked decrease in the concentration of nitrate in the plants because of the conversion of a large proportion to reduced forms. However, the nitrate never disappeared completely from the plants and the rate of reduction was much slower than when the supply of nitrate was maintained. (Fig. 4.17). These results, and others, suggest that the stored nitrate is not all

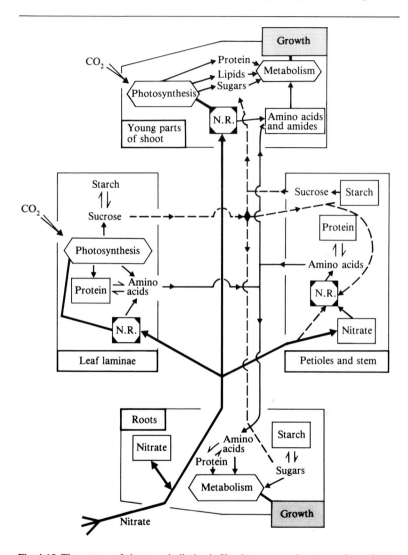

Fig. 4.15 The pattern of nitrate assimilation in *Xanthium pennsylvanicum* where nitrate reductase [NR] is restricted to the shoot nitrate can be stored in both roots and shoot. The transport pathways are indicated by the same symbols as in Fig. 4.14 (after W. Wallace and J. S. Pate 1967).

equally available for reduction and that some is allocated to a less readily available compartment.

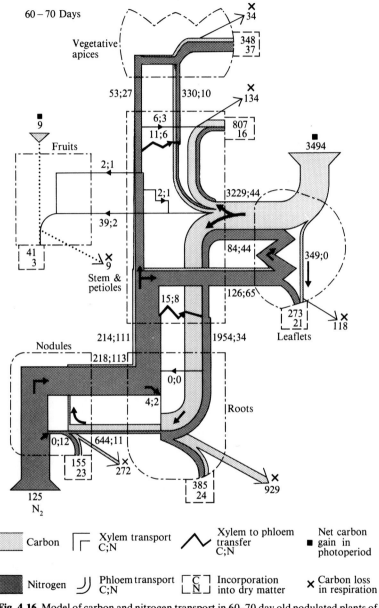

Fig. 4.16 Model of carbon and nitrogen transport in 60–70 day old nodulated plants of *Lupinus albus* relying on nitrogen fixed in nodules. The amounts transported and utilized are given as mg C or N per plant. The values for C appear above or to the left of those for N (after J. S. Pate, D. B. Layzell and D. L. McNeill, 1979).

Table 4.7 Predicted properties of the system transporting carbon and nitrogen in nodulated plants of *Lupinus albus* between days 40 and 80. (After J. S. Pate, D. B. Layzell and McNeil, 1979)

			Time (days after sowing)			
		40–50	50–60	60–70	70–80	40–80
Percentage of intake by vegetative apices taking place through xylem	C	16	15	14	19	15
	N	79	75	73	79	76
Percentage of intake by fruit taking place through xylem	C	–	4	6	9	7
	N	–	30	38	58	53
Percentage of C translocated to nodule returned to shoot as products of N fixation		44	43	34	32	36
Percentage of N incorporated into nodules which is translocated from shoot		36	59	48	61	51
Percentage of N entering vegetative apices in phloem and transferred from xylem in stem and petioles		50	53	60	66	60
Percentage of N entering fruit in phloem transferred from xylem in stem and petioles		–	72	70	75	71
Percentage of N entering leaves in xylem subsequently exported in phloem		47	56	68	57	58

137

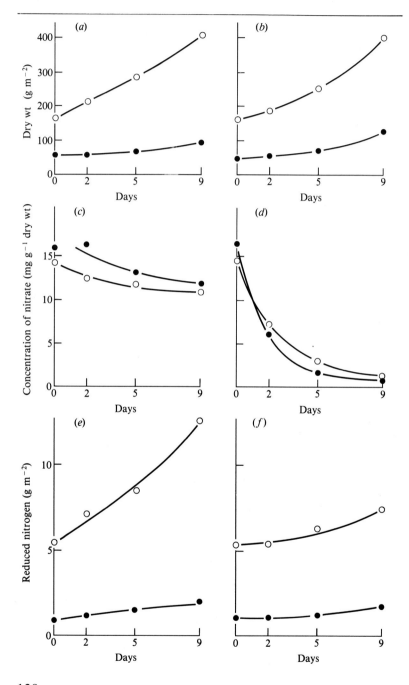

4.3.2 Transport and the growth of stems

There have been few studies of the interrelationships between the supply of assimilates and mineral nutrients to stems and the growth of the latter. (For present purposes potato tubers will be considered with fruit and storage organs.) Probably the best understood system is that in cereals. Here, and in most other situations, two phases can be recognised; the growth of the stem itself, and its use as a temporary store for material used later in the growth of new stems or other organs.

Wheat leaves tend to supply assimilates mainly to the internodes on either side of their node of insertion, with smaller amounts going to some of the nearer higher internodes (Fig. 4.18). Similar distribution patterns seem to be found in other cereals. During internode elongation a considerable number of sieve elements are destroyed but they are formed continually, and the total number present increases throughout growth. This turnover of sieve elements does not seem to lead to any major impediment to translocation and the area available for transport seems to be sufficient for the demands made upon it.

In Chapter 3 the short-term exchange of assimilates between the sieve elements and surrounding tissues was discussed. This exchange can be found in all stems, but is very obvious in species such as sugar cane where the stem is the major storage organ. In other species the stem can act as an intermediate sink which, if necessary, can store materials for long periods before and during the growth of storage organs. These stem reserves can be called on during periods of shortage or maximum demand.

For example, most cereals tend to produce large numbers of tillers. Tiller numbers are increased if the supplies of assimilates and mineral nutrients are plentiful, e.g. at high irradiances or when supplies of fertilizers are abundant. Once the leaves on a tiller start to emerge the import of assimilates from the remainder of the plant stops and the tiller becomes self-sufficient and eventually flowers. However, unless the flowers on the main stem and early-formed tillers are damaged, the later-formed tillers rarely produce flowers and set grain. Instead, they often senesce rapidly and the assimilates and mineral nutrients they contain are transferred to the developing grain. The effects of this transfer on grain growth will be discussed in section 4.3.4. For the present, it is sufficient to note that the tillers can act as temporary stores with the potential for continued growth if necessary.

◀ **Fig. 4.17** The dry weight, nitrate concentration and amount of reduced nitrogen present in six-week-old *Lolium perenne* plants grown in a flowing nutrient solution maintained at $0.1 \ mg \ l^{-1}$ of nitrate throughout the experiment (a,c,e) or reduced to zero for a further 9 days (b,d,f). Open and closed circles refer to shoots and roots respectively (after C. R. Clement, L. H. P. Jones and M. J. Hopper, 1979).

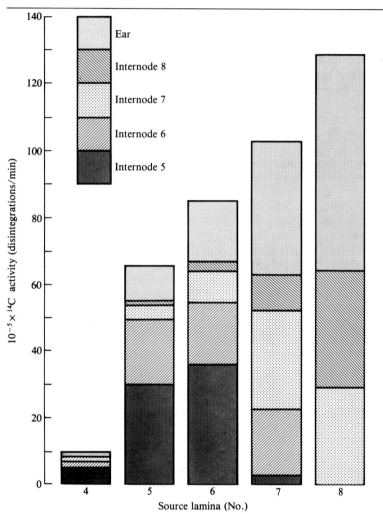

Fig. 4.18 The amounts of ^{14}C-labelled assimilates transported between 22 – 48 days after sowing to internodes 5–8 and the ear of wheat plants after supplying successive leaves with $^{14}CO_2$ (after J. W. Patrick, 1972 *b*).

Teleologically, they are an insurance for the plant against the possibility of damage to the main stem, the early-formed tillers, and the ears which these bear.

Similar patterns of temporary storage in the stem and subsequent transfer to storage organs or fruit can be seen in other species such as potato (see p. 143) and lupin. In lupin the phloem stream moving to

140

the fruit accumulates considerable amounts of N, K and Mg as it moves through the stem, but is is not yet clear whether these are recently absorbed ions moving from the xylem or have been stored for periods in the stem.

The transport of materials to stems and its storage there over much longer periods can be seen in many perennials. In trees and other woody plants the reserves are stored in the ray cells of the xylem and are mobilized and used to support growth in the following spring, before and during leaf expansion. Similarly, the ability of perennial grasses and other pasture plants to form new tillers in the spring, or following grazing by animals or mowing, is dependent mainly on the mobilization of material stored in the basal regions of the plants. This ability is the basis of the success of the grasses in these situations. Some grasses are highly specialized in this regard and have a normal sucrose ⇌ starch interconversion in the leaves during growth but store assimilates for long periods as fructosans.

4.3.3 Transport and the growth of roots

Although the growth of roots is obviously dependent on a supply of assimilates from the shoots there have been very few investigations of this dependence because of the difficulties of studying the growth of underground organs. The evidence available tends to confirm the suspicion that the roots have to compete for the available assimilates and that they do not always compete very successfully with fruit or shoots, especially when assimilates are in short supply. Hence, excavation of field-grown plants suggests that root extension tends to stop when fruit growth starts. Similarly, the root growth of pasture plants is inhibited following defoliation and only restarts when new shoot growth is re-established.

More detailed studies using tomato plants grown in nutrient solutions indicate that in these indeterminate plants there is a turnover of the root system with death of existing roots and the production of new. When fruiting starts the proportion of the total dry matter allocated to the root system falls to a lower, but relatively constant level. An analagous situation can be seen in indeterminate plants with large storage roots such as sugar beet. These tend to have a constant ratio of dry weight of shoot: dry weight of root over much of their period of growth if conditions are reasonably uniform. If the supply of assimilates is increased, for example by raising the irradiance, the plant grows faster but maintains the same root: shoot ratio. If, however, the temperature is changed a new root: shoot ratio is established.

The relative constancy of the root: shoot relation leads to the well-known allometric relationship $R = aS^b$ where R and S are the root and shoot dry weights and a and b are constants. Also, it has made feasible the use of relatively simple mathematical relationships to simulate root growth. There is still, however, little understanding of why some leaves export assimilates to the roots and others to younger leaves, stem apices and fruit. All that can be said is that the roots tend to be supplied by the older leaves, i.e. the sources nearest to them. This level of detail is never achieved in even the most sophisticated models, which usually allocate material to the various sinks from a central pool according to an empirically determined list of priorities.

The models of root: shoot partitioning referred to above are concerned with the dry weights of the organs. Most make some allowances for respiratory losses, but none take account of the exudation of large amounts of material by many roots. These exudates can contain up to 20 per cent of the currently produced assimilates. They must be a valuable source of nutrients for the rhizosphere micro-flora but their value to the higher plant, if any, is unclear. They must, however, place a significant strain on the supply of assimilates to the roots, and more information on their origin and response to environmental factors would be valuable. Their omission certainly emphasizes the empirical nature of most models of root growth.

Species such as sugar beet and carrot with large storage roots also produce a system of much smaller fibrous roots which is concerned with the uptake of water and mineral nutrients. The control of the supply of assimilates to these two types of root, and the pathways followed by mineral nutrients as they move to the aerial parts, have not been widely studied but are of obvious importance. Recent work suggests that the activity of acid invertase in the outer meristematic tissues of sugar beet tap roots controls the concentration of sucrose and influences the initiation of the fibrous roots. In addition, the sucrose synthetase in tap roots seems to be more dependent on the level of uridine diphosphate than adenosine diphosphate whereas the reverse is found in the fibrous roots. It is thought that this difference might be involved in the different rates of sucrose storage in the two types of root. After absorption by the fibrous roots mineral nutrients must have to traverse the massive sink of the tap root before they enter the shoot. The situation is analogous to that in the basal meristems of the leaves of monocotyledonous plants described on p. 125 and the same question requires an answer. How can material traverse an active sink without being removed from the transport system?

4.3.4 Transport and the growth of storage organs and fruit
The effect of transport processes on the growth of storage organs and fruit will be illustrated by a discussion of the situations in potato tubers, cereal grain and the fruit of legumes. Other points of interest which have arisen from investigations of a wide range of other species and organs will be concerned in a final section.

4.3.4.1 Transport and the growth of potato tubers Potato tubers are the swollen internodes produced in the apical regions of underground stems, the stolons. Tuber initiation is an ill-defined event usually taken as the onset of radial swelling about five internodes behind the apex of the stolon which is accompanied by the production of large amounts of starch. The crucial event seems to be the onset of radial rather than extension growth and this results from both cell division and cell enlargement. The apical bud of the new tuber, i.e. the previous stolon apex, continues to produce more internodes which in turn start to swell.

Once initiated the tubers form the major sink for assimilates and mineral nutrients and this leads to major changes in the distribution patterns of assimilates. In one experiment, for example 45.7 per cent of the $^{14}CO_2$ assimilated by untuberized plants in 2 h was exported from the leaves to the rest of the plant in the subsequent 18 h. After tuberization 71.7 per cent was exported. How this increase is achieved is not known. It cannot be attributed to a major increase in the absolute growth rate of the stolon/tuber system since this must be preceded by an increased supply of assimilates. Various investigations have suggested that tuber initiation and growth can have a feed-back effect which tends to increase the rate of photosynthesis and similar effects have also been seen in other species.

The rate of growth of the crop of tubers seems to be determined soon after tuber initiation and remains virtually constant throughout most of the period of tuber growth unless there is a major environmental restriction such as drought. This constancy is dependent, in part, on the transfer of stored material from the stems to compensate for short-term fluctuations in the supply of current assimilates. This transfer of stored material tends to become more important towards the end of tuber growth when it can account for 25 per cent of the material entering the tubers. Attempts have been made, but with little success, to try and relate the growth of individual tubers to the supply of assimilates from specific leaves or groups of leaves. Rather, the tubers compete for the available assimilates, the maximum size attained by the tubers being a positive function of the size of the photosynthetic surface supplying them and an inverse function of the number of tubers.

If there is a major reduction in the supply of assimilates, say during a period of water deficit, radial growth may stop but the apex continues to grow as a stolon until adequate supplies of assimilates are available again. A new, more distal, tuber is then formed. It is unusual for the growth of the first-formed tuber to resume, and in some instances the starch it contains can by hydrolysed and the sugars transferred to the new tuber. This sequence of events can be repeated several times producing 'chains' of tubers on a single stolon. The control processes which govern the movement of assimilates through the phloem of the primary tuber and into the second tuber are not known but may be related to the observation that sucrose synthetase, an enzyme closely involved in starch synthesis in the tuber, appears to be associated with the continued supply of assimilates. When the latter declines so also does the activity of the enzyme and since it appears not to be reactivated the primary tuber no longer constitutes a sink.

Because a potato tuber is a large sink connected to the source of assimilates by a restricted transport pathway, the stolon, tubers have been used on several occasions as model systems to examine the rate of translocation. These experiments have produced values of the correct order of magnitude, but have to be treated with caution. The anatomy of stolons is complicated by the shortness of the internodes, many anastomoses of the vascular bundles, and the presence of files of sieve tubes which connect the vascular bundles in the internodes. It is, therefore, difficult to define the cross-sectional area of the transport pathway. Further, some estimates of the amount of material transported were based simply on the final weight of the tuber with no allowance being made for the CO_2 lost by respiration of the tuber. Depending on the temperature, this could easily lead to an error of 25–30 per cent.

The transport of ions and water into tubers is rather more complicated than that of assimilates. Tubers grow underground in conditions which do not usually favour the loss of water. Moreover, they contain large amounts of water (usually about 80 per cent of the fresh weight) and are connected by the xylem to the leaves which can transpire readily. Hence, when the leaves start to lose water, the increased gradient in water potential between the leaves and tubers causes water to be withdrawn from the tubers. The relative amounts moving to the leaves from the roots and tubers depend on the resistances to flow in the two paths but tubers can lose up to about 10 per cent of their fresh weight when transpiration starts and this is only regained when transpiration stops (Fig. 4.19).

This reversal of the direction of flow through the xylem of the stolons restricts the delivery of ions to the tubers to periods when there is no transpiration e.g. darkness. Hence, calcium, and other ions such as strontium, which do not move readily through the phloem cannot

Fig. 4.19 (*a*) Trace from a recording balance showing the changes in fresh weight of a ▶ potato tuber which take place when the lights of the controlled environment room were switched on and off. NB the total weight shown on the ordinate was the maximum change in weight which could be recorded, not the total weight of the tuber.
(*b*) Ratemeter trace showing the lack of any effect of changes in transpiration on the movement of ^{32}P through a potato stolon.
(*c*) Similar to (*b*) but showing the effect of transpiration on the movement of ^{89}Sr through a potato stolon (after D. A. Baker and J. Moorby, 1969).

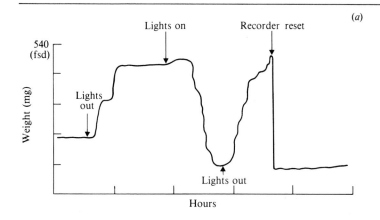

(a)

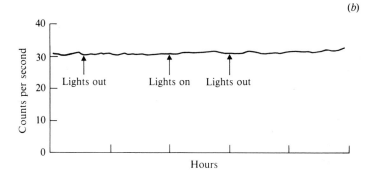

(b)

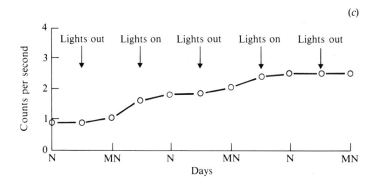

(c)

enter the tubers when the leaves are transpiring. In contrast, the phloem-mobile ions such as phosphates, can enter the tubers with the assimilates moving from the leaves. As this transfer continues it can lead to an eventual decline in the concentration of the ions in the shoots and senescence of the photosynthetic surface.

4.3.4.2 Transport and the growth of cereal grains The cereal grains form the most important group of crops and because of this have been extensively studied. For the sake of brevity, most of this section will be concerned only with a description of transport into and growth of wheat grains but, where necessary, reference will be made to other species.

All the cereal grains have been changed markedly by selection from wild ancestors and by active breeding programmes and this has been accompanied by profound changes in the transport systems of the crops. For example, the harvest index, i.e. the proportion of the total dry matter allocated to the grain, has increased from about 30 per cent in old varieties of wheat, barley, maize and rice used before the introduction of modern breeding programmes to about 50–55 per cent in modern dwarf varieties. There has, therefore, been an increase in the proportion of assimilates translocated to the grain. In wheat this has been accompanied by an increase in the cross-sectional area of the phloem in the peduncle in order to accommodate the extra material moving through it. This results in a linear relationship between the estimated maximum rate of assimilate transport into the ear and the area of phloem through which this occurs (Fig. 4.20). Further, if the grain weight per ear is reduced by increasing the period of vernalization it is accompanied by a reduction in the area of phloem in the peduncle.

These relationships show how the overall pattern of translocation can be changed but understanding how the transport systems operate requires knowledge of the developmental processes in each species. In wheat, for example, after the stem apex has produced a number of leaves it starts to elongate and the subsequent primordia develop into the spikelets of the ear. These, in turn, produce smaller primordia which form the individual florets. As in the early development of leaves, there is a period when the meristematic regions have no direct vascular connections with the rest of the plant. They are, therefore, dependent on the diffusion of substrates from the most recently differentiated phloem. Subsequently, better transport pathways are developed, but the vascular bundles of the rachis are not connected to those leading to the grain. They are separated by regions of transfer cells at the base of the spikelet and each of the caryopses and attached glumes (Fig. 4.21). Further, the sizes of the grain in a wheat ear vary systematically. The largest grains are usually found in the spikelets in

146

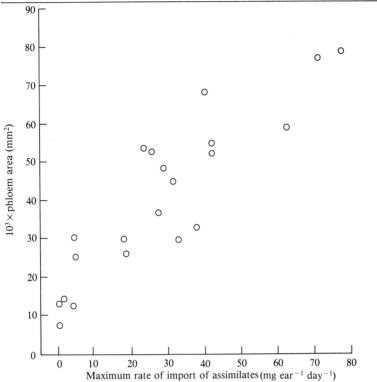

Fig. 4.20 The relationship between the phloem area of the culm and the estimated rates of import of carbohydrates by the ears of a range of wheat species and cultivars (after L. T. Evans, R. L. Dunstone, A. M. Rawson and R. F. Williams, 1970).

the middle part of the ear but there are also differences within each spikelet. The final weight, and growth rate, of the second or third grains on a spikelet are usually greater than the first, and the rates of growth of the more distal grains are even slower.

The reasons for the different growth rates of grains are not known. The concentration of sucrose in the leaves of wheat shows diurnal fluctuations corresponding to the rates of photosynthesis. In contrast, the concentration of sucrose in the grain shows only minor variations suggesting that the transfer into the grain is limited in the final stages of the transport process. The discontinuities in the phloem at the bases of the spikelets could provide such a limitation, but this is not proven. Attempts to simulate the situation have used a model in which the ear is represented by a complex network of varying resistances supplied from a single source, for example, the transfer cells in the rachis. The behaviour of the model suggests that there are probably differences in

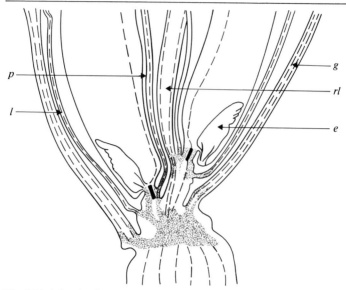

Fig. 4.21 A longitudinal section through a wheat spikelet showing the vascular tissues (broken lines) and associated transfer cells (stippled areas) and the sterile glumes, g, paleas p, rachilla rl, lemmas l and embryos e (after S. Y. Zee and T. P. O'Brien, 1971).

the growth potentials of the individual grains, in addition to those imposed by the various transport resistances, but the nature of these differences requires further investigation.

There are three possible sources for the dry matter which accumulates in the grain; material assimilated or absorbed during the period of grain growth and transported immediately from the leaves or roots, material accumulated before anthesis and stored in stems and sterile tillers, and assimilates produced by photosynthesis within the developing ear. The relative contributions of the various sources to the final grain yield vary with the species and conditions during growth.

Most of the carbon accumulated by wheat grains is assimilated by the flag leaf. The transport of material assimilated before anthesis seems to be used to buffer the supply to the grain against any short-term fluctuations. The absolute amount of this material which is transported seems to be fairly stable and in wheat and other cereals typically contributes about 10 per cent of the final grain yield, but in barley it has been shown to account for up to 70 per cent when post-anthesis conditions were very unfavourable to photosynthesis. The dynamics of the assimilation, storage and transport of assimilated carbon in wheat can be seen in Fig. 4.22. The difference between the carbon lost from the leaves and stems and that accumulated in the

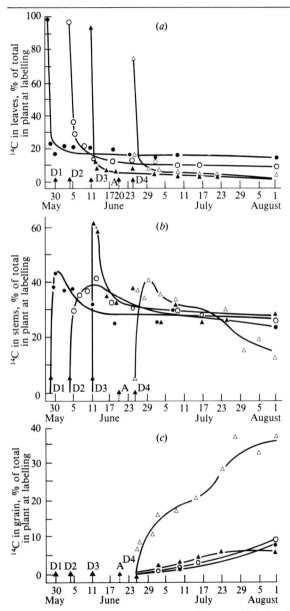

Fig. 4.22 Carbon-14 present in the leaves (*a*), stems (*b*) and grain (*c*) of wheat plants at various times after exposure of the plants to $^{14}CO_2$. Seed sown on 1.12.74 (Cambridge, UK) and exposed to $^{14}CO_2$ on 28.5.75 (●——●), 4.6.75 (○——○), 11.6.75 (▲——▲) and 25.6.75 (△——△). The time of anthesis (A) is indicated (after R. B. Austin, J. A. Edrich, H. A. Ford and R. D. Blackwell, 1977).

grain can be attributed to respiratory losses and transport to the root system.

Photosynthesis by the ear in barley is more important than in wheat and can provide from 10–20 per cent of the final grain dry weight. This greater contribution is because the awns in barley increase the amount of radiation intercepted by the ear. A similar effect can be seen in the awned varieties of wheat.

The major sources of assimilates in maize are the leaves which subtend the cobs, i.e. the leaves closest to the sinks. Maize leaves do not have such a pronounced sequential senescence as those of wheat and barley and the older leaves are able, therefore, to supply substantial amounts of assimilates to the developing grain even though they intercept less radiation than the leaves at the top of the canopy. The senescence of rice leaves is also relatively slow. Hence, even though the rice pannicle is terminal, the role of the flag leaf as a major source of assimilates is not as pronounced as in wheat and barley and a significant amount of material is transported from more distant leaves.

The interrelations between the transport to the grain of currently absorbed nitrogen, and nitrogen stored in the stems and senescing leaves, is analogous to the situation with assimilates decribed above. The prevention of uptake after anthesis seems to promote the transfer of nitrogenous materials from the older leaves and, in one experiment, led to an increase in the concentration of nitrogen in the grain. The transport of other phloem–mobile ions, e.g. phosphate, into the grain seems to follow similar patterns (Fig. 4.23).

4.3.4.3 Transport and the growth of fruit Early work with large fruiting tomato plants suggested that the carbohydrates exported by the leaves formed a central pool which could be tapped by the developing fruit. More recent work has shown that, as in most other species, the exported carbohydrates tend to move into the nearest fruit. However, the transport system is very responsive and the loss of a leaf has little effect on the final weight of the subtended fruit.

The same situation can be found in peas and other legumes, but, where the leaves are opposite, the arrangement of the vascular system causes most of the carbohydrate to be supplied from the leaves immediately above and below the fruit. There is little transfer from leaves on the opposite side of the stem. The vascular bundles in tomato are bicollateral and there is some evidence that the outer phloem is concerned mainly with transport downwards, whereas upward transport tends to be confined to the inner phloem.

The rate of transport of assimilates into tomato fruit (mg carbon d^{-1}) appears to be proportional to the sucrose concentration in fruit (Fig. 4.24) and the situation can be described by an equation analogous to that outlined on pp. 33. The rate gradually decreases to

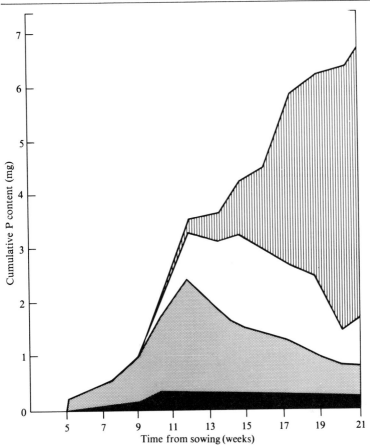

Fig. 4.23 The cumulative phosphorus content of the roots ■ , leaves ▨ , stem □ and ear ▥ of field-grown wheat plants. Equal amounts of phosphate and potassium (75 kg ha^{-1}) were applied before planting and nitrogen (37.5 kg ha^{-1}) applied on two occasions 4 and 8 weeks after sowing (after G. E. S. Mohamed and C. Marshall, 1979).

about half the initial value as the fruits approach their final size. In large fruit which are cooled the direction of transport can be reversed and the fruit can be induced to export carbon (Fig. 4.25 and Table 4.5) thus emphasizing a point, made on several occasions, that there are no absolute sources or sinks. Other experiments have shown that over a 24 h period the rate of export from a leaf to a fruit is almost independent of the rate of photosynthesis in the leaf. Any deficit in current photosynthesis is made up by the export of stored materials and any surplus is stored (Table 4.8). It can also be seen that in plants

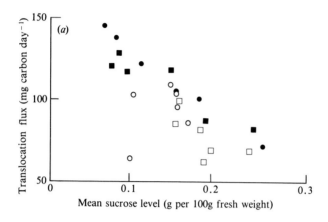

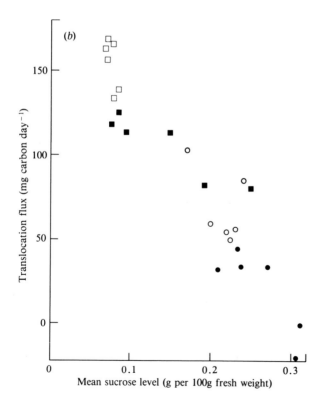

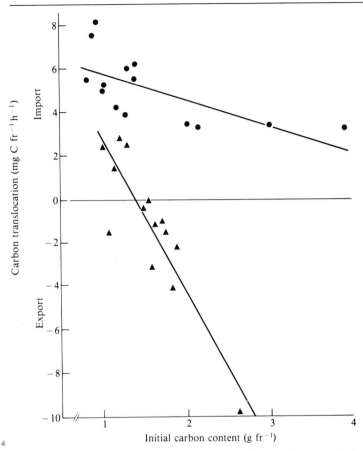

Fig. 4.25 The rate of carbon translocation into or out of tomato fruit over a 48 h period as a function of the initial carbon contents of fruit maintained at 25 ± 4 °C (•——•) or cooled to 5 ± 1 °C (▲——▲) (after A. J. Walker and L. C. Ho, 1977).

◄ **Fig. 4.24** The relationship between the rate of translocation into tomato fruit over a 48 h period and the mean concentration of sucrose in the fruit throughout the period. (*a*) At 25 °C and a range of mean fruit sizes 386.6 mg carbon (•), 931.0 mg carbon (○), 159.1 mg carbon (■) and 2401.7 mg carbon (□) and (*b*) similarly sized fruit at 7.5 °C (•), 17.5 °C (○), 25.0 °C (■) and 37.5 °C (□). NB. The sucrose concentration in the rest of the plant was essentially constant (0.47 ± 0.06 g sucrose to 100 g fresh weight) and hence as the sucrose concentration in the fruit declined the gradient along the transport pathway increased (after A. J. Walker and J. H. M. Thornley, 1977).

Table 4.8 The carbon metabolism of leaves and fruit on tomato plants kept at various irradiances for 24 h after pruning the plants to one leaf and one fruit (after L. C. Ho, 1979).

Expt	Light flux density ($W\ m^{-2}$)	Integrated radiation ($MJ\ m^{-2}$)	Leaf area (dm^2)	Carbon fixed (mg)	Carbon exported from leaf (mg)	Fruit volume (cm^3)	Carbon imported by fruit (mg)	Carbon respired by fruit (mg)
1	0.5	0.04	3.1 ± 0.27	2.7 ± 0.8	66.4 ± 7.2	18 ± 4.0	84.3 ± 13.4	15.0 ± 0.9
2	10	0.86	2.8 ± 0.41	46.7 ± 1.9	77.5 ± 7.2	15 ± 4.0	93.8 ± 15.4	15.8 ± 1.2
3	20	1.73	3.2 ± 0.39	111.1 ± 6.7	107.2 ± 2.9	18 ± 5.1	103.0 ± 19.4	16.8 ± 1.1
4	40	3.46	2.7 ± 0.42	178.4 ± 6.6	126.1 ± 4.2	12 ± 4.8	102.0 ± 25.4	17.2 ± 0.9
5	70	6.91	2.6 ± 0.41	217.6 ± 11.8	119.9 ± 8.2	13 ± 3.9	94.7 ± 13.1	18.7 ± 1.8
6	100	8.64	3.3 ± 0.58	293.8 ± 24.9	123.2 ± 8.4	19 ± 3.8	119.9 ± 16.2	19.3 ± 0.9

grown at low irradiances the rate of import of carbon into the fruit is greater than the rate of export from the leaves and the reverse is found at high irradiances. These results suggest that even in these plants, which were pruned to one leaf and a single subtended fruit, there was sufficient reserve capacity in the intervening stem and petiole to help buffer the transport system against the treatments imposed.

A significant proportion of the carbon transported to a fruit is lost by respiration (Table 4.8), and this respiration rate tends to decrease as fruits grow but then increases rapidly in the climacteric at ripening. The situation in legume fruits is less straight-forward than in tomatoes because of the more complex structure of the fruit. Most of the carbon used in growth is translocated into the fruit from the leaves and the direct uptake of CO_2 by the pods and seeds makes little contribution to the final dry weight of the seeds. However, considerable amounts, about 50–70 per cent in soyabeans, of the CO_2 lost by respiration of the pods and seeds is refixed by photosynthesis in these tissues. This uptake is aided by the presence in pea cotyledons of significant amounts of PEP carboxylase. This permits CO_2 fixation in both light and darkness and would help, therefore, to maintain very low concentrations of CO_2 in the pod.

A more complete analysis of the turnover of carbon, nitrogen and water in lupin fruits can be seen in Fig. 4.26. Transport through the phloem supplies 98 per cent of the carbon, 89 per cent of the nitrogen and 40 per cent of the water which enters the fruit, this latter becoming relatively more important as the seeds start to swell. As the fruits mature 80 per cent of the nitrogen in the pod is transferred to the seeds and the pod and seed tissues lose 93 per cent and 25 per cent of their water respectively.

Much of the nitrogen supplied to legume fruit through the phloem comes from the leaves. It has been suggested that in some species, e.g. soybean, this in turn causes early leaf senescence and death of the plant so imposing a limitation on the ultimate seed yield. If leaf senescence in soybean is delayed by the application of a mixture of a cytokinin and an auxin, photosynthesis and nitrogen uptake continue for a longer period and there are no adverse effects on pod growth. It is not yet clear, however, whether such treatments can be applied commercially.

4.3.5 Transport out of seeds and storage organs

Previous sections of this chapter have considered the movement of materials into seeds, fruit and storage organs as though this was an end in itself. It is, however, merely an intermediate step in the production of another generation of plants. The initial growth of the new generation depends on the mobilization of the reserves laid down in the seeds, etc. and their transfer to the new plants.

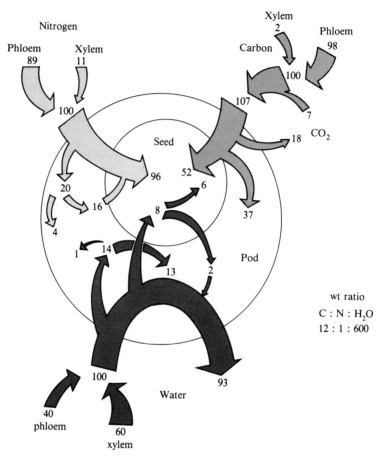

Fig. 4.26 The movement of carbon, nitrogen and water into and out of the fruit of white lupin throughout the whole period of growth. The results are expressed relative to a net intake through the vascular tissue of 100 units by weight of C, N or H_2O, the actual intake varied in the ratio 12 : 1 : 600. Note that there was an additional intake of 7 units of C by photosynthesis of the fruit (after J. S. Pate, P. J. Sharkey and C. A. Atkins, 1977).

Seeds, and the buds on tubers, have varying periods of dormancy, and the breaking of the latter is accompanied by the production of gibberellins in the embryos and buds. The gibberellins in turn stimulate the activity of hydrolytic enzymes and the action of enzymes such as invertase leads to the conversion of the storage materials into forms which can be transported to the developing meristems and later the new roots and shoots.

In most instances a single plant develops from each seed. The situation is rather different in potato tubers where each bud can produce a sprout, and eventually an independent plant, but the early growth of each bud is dependent on its ability to compete with the other buds for the reserves of the mother tuber. There is sufficient phloem in the tuber to ensure that all the buds can compete on equal terms for access to all the tuber reserves and they are not restricted to those in the regions immediately adjacent to the buds. In addition to this competition, an apical dominance system is established which can limit the number of buds which grow and eventually emerge.

This early transport of reserves to the developing roots and shoots is

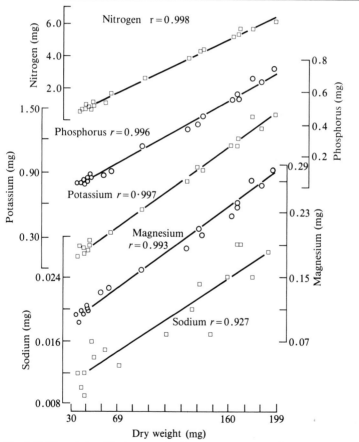

Fig. 4.27 The relationships between the dry matter contents and amounts of N, P, K, Mg and Na in the cotyledons of germinating pea seeds as material is moved from the cotyledons over a period of 4 weeks (after O. D. G. Collins and J. F. Sutcliffe, 1977).

very important if the plant is to become established and able to compete successfully with its neighbours. Small differences in the initial size and early relative growth rate of seedlings tend to persist, and can be maintained throughout the development of the plant.

The mineral nutrient composition of seeds is remarkably constant. Even when the mother plants are grown in a wide range of nutrient concentrations it is rare for the concentrations in the seeds to vary by more than a factor of two. This constancy extends to the material transported from the seeds to the developing plants.

For example, the rates of transport of a range of ions relative to dry matter from the cotyledons of pea seeds remained constant for at least four weeks, each 100 mg of dry matter containing 0.8, 0.1, 3.4, 0.01 and 0.5 mg of K, Mg, N, Na and P respectively (Fig. 4.27). When the amounts moved in these four weeks are expressed as percentages of the original amounts of nutrients in the seed they show differing absolute rates of movement, but if expressed as percentages of the total amounts of each ion moved they fall on a single straight line (Fig. 4.28) thus suggesting that ions are transported in constant proportion to each other but that differing amounts of the ions present are available for transport. This constancy is not always found, and in oat seed the rates of transport of ions from the endosperm show curvilinear relationships with the rate of transport of dry matter (Fig. 4.29).

The behaviour of calcium is once again atypical in some species. In peas, for example, more than half of the calcium in the seeds is found in

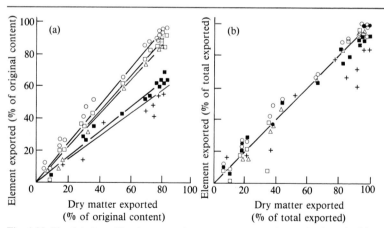

Fig. 4.28 The data from Fig. 4 expressed as amounts exported over the 4 weeks (*a*) as percentages of the original contents and (*b*) as percentages of the amounts transported after 4 weeks. N ($\triangle$); K ($\square$); P ($\bigcirc$); Mg ($\blacksquare$); Na (+) (after D. D. G. Collins and J. F. Sutcliffe, 1977).

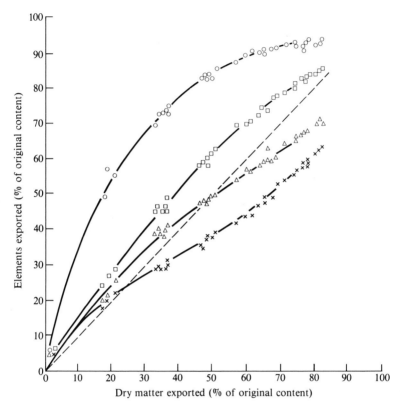

Fig. 4.29 The relationships between the export of dry matter and N, P, K, and Mg from the endosperm of etiolated oat seedlings during the first 7 days of germination in 10^{-4} M Ca Cl₂. The lines are fitted quartic polynomials. N (□); P (△); K (○) and Mg (×) (after Q. A. Baset and J. F. Sutcliffe, 1975).

the testa and is not transported to the developing embryo. Further, only a small proportion of the remaining cotyledonary calcium is transferred. Similar circumstances are probably present in other species. Hence, symptoms of calcium deficiency can often be found when potatoes are sprouted for long periods prior to planting, or in the production of bean sprouts. In both situations germination takes place over a long period in conditions which do not favour transpiration or the uptake of external calcium.

159

4.4 The movement of exotic substances

Much of modern agriculture depends on the use of pesticides and herbicides which are applied either to the soil or sprayed onto the plants. Both methods tend to be wasteful, and can be potential causes of risk. Typically, 20–50 times the amount of material necessary has to be applied to ensure that sufficient arrives at the sites of action.

Several techniques are used to try and reduce the amounts of active ingredients which have to be applied and similar general considerations apply irrespective of the type of compounds involved. Before absorption, materials which are applied to the soil have to move to the roots in the soil solution, or the roots have to grow into the treated regions of the soil. Thereafter, the uptake and transport of pesticides are similar to those of nutrient ions or non-polar materials and are governed by factors similar to those outlined in Chapter 3. The present discussion will be concerned only with materials applied to the foliage of plants.

The aim of any pesticide application is to ensure that an adequate amount of the active ingredient is delivered to the target site in the most efficient manner. The techniques used to ensure this can be divided into three classes:

(a) Means of ensuring that the maximum possible proportion of the chemical sprayed is delivered to the target plant or animal.
(b) Ways of facilitating the absorption of material after deposition.
(c) Attempts to ensure that the active ingredient can move to the site of action after penetration.

The delivery to the target of a large proportion of the material sprayed is often a function of droplet size and spray pattern and the physical principles governing these factors are reasonably well understood and need not concern us. However, a recent development which is of interest is the use of techniques to produce droplets or aerosols with a positive electrical charge which are attracted to the negatively charged plant surfaces.

The application methods used are closely related to the formulation of the pesticide i.e. the use of various surface active agents and additives to control factor such as surface tension, lipid or water solubility, the ability to form emulsions and the rate of drying of the drops. The aim is to ensure that when the droplets hit the plant they stick to it, and do not bounce off, and that they stay there long enough to allow the pesticide to penetrate the very hydrophobic cuticle. Foliar uptake is not fully understood. It is clear that the surface tension of the spray is usually too high to permit penetration of the stomatal pores even when they are fully open. Nevertheless, entry often takes place near the stomata; possibly because of changes in the nature of the cuticle or defects caused by flexing of the guard cells.

After penetration, the pattern of movement is governed by the nature of the material. Some substances move only through the xylem whereas others can move through the phloem, i.e. the two groups make use of apoplastic and symplastic pathways respectively. Because of its obvious practical importance, many attempts have been made to relate phloem mobility to chemical structure but there are still no generally accepted rules. The possession of a weakly acidic group seems to aid penetration of the symplast and movement through the phloem, but this is not invariably true. Further, very easy penetration is often accompanied by an equally easy loss from the sieve tubes to the surrounding tissues thus resulting in only limited transport through the phloem. If we are to reduce the amounts of pesticides sprayed on to plants, and the attendant risks of pollution, it is very important to understand the factors which control the foliar uptake of exotic substances and their movement through the sieve tubes so that it becomes possible to synthesize materials with known patterns of mobility.

Further reading and references

Further reading
Most of the books referred to at the end of previous chapters have sections which are relevant to this chapter. So also do most issues of the *Annual Review of Plant Physiology*.

Other references
AUSTIN, R. B., EDRICH, J. A., FORD, M. A. and BLACKWELL, R. D. (1977) The fate of dry matter, carbohydrates and ^{14}C lost from the leaves and stems of wheat during grain filling, *Ann. Bot.*, **41**, 1309–21.
BAKER, D. A. (1969) Transport pathways in sprouting tubers of the potato, *J. exp. Bot.*, **20**, 336–340.
BAKER, D. A. and MOORBY, J. (1969) The transport of sugar, water and ions into developing potato tubers, *Ann. Bot.*, **33**, 729–741.
CLARKSON, D. T., SHORE, M. G. T. and WOOD, A. V. (1974) The effect of pretreatment temperature on the exudation of xylem sap by detached barley root systems, *Planta*, **121**, 81–92.
CLEMENT, C. R., JONES, L. H. P. and HOPPER, M. J. (1979) Uptake of nitrogen from flowing nutrient solution: effect of terminated and intermittant nitrate supplies, pp. 123–133, in Hewitt E. J. and Cutting C. V. (eds), Nitrogen assimilation of plants. Academic Press, London.
COLLINS, O. D. G. and SUTCLIFFE, J. F. (1977) The relationship between transport of individual elements and dry matter from the cotyledons of *Pisum sativum* L., *Ann. Bot.*, **41**, 163–171.
EVANS, L. T., DUNSTON, R. L., RAWSON, H. M. and WILLIAMS, R. F. (1970) The phloem of the wheat stem in relation to requirements for assimilated by the ear, *Aust. J. biol. Sci.*, **23**, 743–752.
HANNAM, R. V. (1968) Leaf growth and development in the young tobacco plant, *Aust. J. biol. Sci.*, **21**, 855–870.

HO, L. C. (1979) Regulation of assimilate translocation between leaves and fruits in the tomato, *Ann. Bot.*, **43**, 437–448.

HOPKINSON, J. M. (1964) Studies on the expansion of the leaf surface IV. The carbon and phosphorus economy of the leaf, *J. exp. Bot.*, **15**, 125–137.

LUSH, W. M. (1976) Leaf structure and translocation of dry matter in a C_3 and a C_4 grass, *Planta*, **130**, 235–244.

MOHAMED, G. E. S. and MARSHALL, C. (1979) The pattern of distribution of phosphorus and dry matter with time in spring wheat. *Ann. Bot.*, **44**, 721–730.

MOORBY, J. (1964) The foliar uptake and translocation of caesium, *J. exp. Bot.*, **15**, 457–469.

MOORBY, J., TROUGHTON, J. H. and CURRIE, B. J. (1974) Investigations of carbon transport in plants II. The effects of light and darkness and sink activity on translocation, *J. exp. Bot.*, **25**, 937–944.

MOORBY, J. and JARMAN, P. D. (1975) The use of compartmental analysis in the study of movement of carbon through leaves, *Planta*, **122**, 155–168.

MUNNS, R. and PEARSON, C. J. (1974) Effect of water deficit on translocation of carbohydrate in *Solanum tuberosum*, *Aust. J. Plant Physiol.*, **1**, 529–537.

PATE, J. S. (1980) Transport and the partitioning of nitrogenous solutes. *Ann. Rev. Pl. Physiol.*, **31**, 313–340.

PATE, J. S., LAYZELL, D. B. and McNEIL, D. L. (1979) Modelling the transport and utilization of carbon and nitrogen in a nodulated legume, *Plant Physiol.*, **63**, 730–737.

PATE, J. S., SHARKEY, P. J. and ATKINS, C. A. (1977) Nutrition of a developing legume fruit. Functional economy in terms of carbon, nitrogen, water, *Plant Physiol.*, **59**, 506–510.

PATRICK, J. W. (1972*a*) Vascular system of the stem of the wheat plant II. Development. *Aust. J. Bot.*, **20**, 65–78.

PATRICK, J. W. (1972*b*) Distribution of assimilate during stem elongation in wheat, *Aust. J. Biol. Sci.*, **25**, 455–67.

PEARSON, C. J. and STEER, B. T. (1977) Daily changes in nitrate uptake and metabolism in *Capsicum annuum*, *Planta*, **137**, 107–112.

SETH, A. K. and WAREING, P. F. (1967) Hormone–directed transport of metabolites and its possible role in plant senescence, *J. exp. Bot.*, **18**, 65–77.

WALKER, A. J. and HO, L. C. (1976) Young tomato fruits induced to export carbon by cooling, *Nature*, **261**, 410–411.

WALKER, A. J. and HO, L. C. (1977) Carbon translocation in the tomato: effects of fruit temperature on carbon metabolism and the rate of translocation, *Ann. Bot.*, **41**, 825–832.

WALKER, A. J. and THORNLEY, J. H. M. (1977) The tomato fruit: import, growth, respiration and carbon metabolism at different fruit sizes and temperature, *Ann. Bot.*, **41**, 977–832.

WALLACE, W. and PATE, J. S. (1967) Nitrate assimilation in higher plants with special reference to the cocklebut (*Xanthium pennsylvanicum* Wallr.). *Ann. Bot.*, **31**, 213–228.

WARDLAW, I. F. (1967) The effect of water stress on translocation in relation to photosynthesis and growth I. Effect during grain development in wheat, *Aust. J. biol. Sci.*, **20**, 25–39.

WARDLAW, I. F. (1976) Assimilate movement in *Lolium* and *Sorghum* leaves I. Irradiance effects on photosynthesis, export and the distribution of assimilates, *Aust. J. Plant Physiol.*, **3**, 377–387.

WATSON, B. T. (1975) The influence of low temperature on the rate of translocation in the phloem of *Salix viminalis* L., *Ann. Bot.*, **39**, 889–900.

WILLIAMS R. F. and WILLIAMS, C. N. (1968) Physiology of growth in the wheat plant IV. Effects of day length and light-energy level, *Aust. J. biol. Sci.*, **21**, 835–54.

ZEE, S. Y. and O'BRIEN, T. P. (1971) Vascular transfer cells in the wheat spikelet, *Aust. J. biol. Sci.*, **24**, 35–49.

162

Chapter 5

Conclusions

It will be apparent that we are still ignorant about many important aspects of the transport systems of plants; for example, how ions enter the xylem, the details of the mechanism(s) by which sugars move through the sieve tubes, why certain groups of cells become sinks or sources and how they change from one to the other.

The mechanism of the movement through the sieve tubes has been of major interest to transport physiologists for over 50 years but surprisingly little progress has been made. There is general agreement that there must be a *flow* of solution through the sieve tubes but there is a wide variety of views on how the flow is generated and how it can be reconciled with the structure of the sieve tubes. The situation is almost analogous to that in economics, 'if two economists agree they are probably both wrong'. I believe that we have exploited present techniques almost to their limits and that we will make little further progress until new techniques become available. It has not been possible to use modern techniques of electron microscopy to decide which mechanisms are anatomically feasible and resolution by these means seems unlikely. It will be necessary, therefore, to rely on more physiological methods to make any advance and the value of any further understanding likely to result from such investigations has to be balanced against the great effort required to make such an advance.

More profitable areas for further investigation would appear to be the initiation and behaviour of sources and sinks, the biochemical changes which take place in them, and how these changes relate to cell expansion and division. Only when information on these topics has been obtained in sufficient detail to permit a quantitative description

163

of the system will we be approaching a sufficient depth of understanding. It should then be possible to give a more coherent account of the material covered in Chapter 4 rather than the very episodic coverage presented there. In addition, we would then be in a position to give a logical quantitative description of the partitioning of assimilates and mineral nutrients during growth rather than having to rely on arbitrary partition functions as at present.

Index

Abies grandis, 23
abscissic acid (Ab A), 110, 121
actin, 70
Alaria marginata, 5, 6, 54
albuminous cells, 11
alkaline earths, 104
allometric relationship, 141
Amaranthus caudatus, 40, 44, 45, 112, 114
aphids, 54, 63, 64, 66
Apium graveolans, 12, 37
apoplasm (apoplast), 28, 35, 36, 37, 72
apple, see *Malus sylvestris*
area of transport system, 127, 128, 144, 146, 147
asparagine, 133
aspartate, 35, 40
aspiration of pits, 22, 24
ATP, 56, 57, 66
autoradiography, 2, 8, 17, 37, 104
Avena sativa, 158, 159

barley, see *Hordeum vulgare*
barrel pores, 18
Beta vulgaris, 35, 60, 72, 141, 142
bicollateral bundles, 19, 129, 150
bidirectional transport through phloem, 13, 65, 66, 67
Brownian movement, 13
buffering
  of translocation, 46, 139, 140, 143, 148
  of water transport, 87

bundle sheath cells, 35, 40, 104

caesium, 105, 106, 115, 116
calcium, 91, 92, 93, 99, 100, 132, 144
  in phloem, 54, 56, 57, 104, 158, 159
callose, 6, 11, 16
cambium, 20, 120
Capsicum, 103, 113
carrot, see *Daucus*
Casparian strip, 87, 89, 90, 91, 93
celery, see *Apium graveolans*
cereals, 47, 139, 146–50
chloroplasts, 34, 35, 41, 57
citrate, 101
climacteric, 155
clubmosses, 18, 19
companion cell, 10, 11, 17, 30
compartmentation of carbohydrates, 41, 44, 45, 73, 112
competition for assimilates and reserves, 141, 143, 157
concentration of ions, effect on uptake, 83, 84, 85, 86
concentration of sieve tube contents, 53–64
  diurnal fluctuations, 57, 58, 59
contractile structures in phloem, 70
cotton, see *Gossypium*
Crassulacean acid metabolism (CAM), 34, 35, 39, 40, 41
Cucumis sativus, 26, 131
Cucurbita maxima, 16

Index

Cucurbita
  melopeepo, 54, 66
  pepo, 90, 94
cuticle, 18

Daucus carota, 142
desmotubule, 28, 29, 31
dictyosome, 7
dictyostele, 19
diffuse porous species, 21
diffusion
  accelerated, 62
  equation, 33
  of CO_2, 33, 34
  coefficient, 33, 62
  into primordia, 123, 146
  through phloem, 62, 65, 70
  through plasmadesma, 36
Digitaria sanguinalis, 35
direction of translocation, 13, 61, 65, 66, 67, 125, 130, 142
discrimination between ions, 79, 90
distribution
  of assimilates, 109, 114, 115, 127, 129, 130, 131, 139, 143, 146, 148, 150
  of ions, 115, 116, 130, 131, 132, 150, 157
dormancy, 156
double layer of electrical charge, 80
Dryopteris, 19

EDTA, 101
electrochemical potential gradient, 81, 82, 84
endodermis, 87, 89, 90
endoplasmic reticulum, 4, 8, 11, 14, 15
energy requirement
  for ion uptake, 82, 84
  for translocation, 66, 67, 69
evolution of transport systems, 1–27
exchange
  between sieve cells and surrounding cells, 48, 49, 50, 54, 63, 72, 106
  between xylem and surrounding cells, 99, 100, 101
  coefficient, 44, 45
  of ions, 81, 100
export from leaves
  of carbohydrates, 40, 41, 45, 46, 59, 109, 125–36, 140, 143
  of mineral nutrients, 130–6

feed-back effects, 143
ferns, 18, 19
fixation of specimens, 13–16

flow
  through plasmodesmata, 36, 90, 91
  through sieve tubes, 49, 53, 60, 61, 63–6, 70
  through xylem, 90
flux
  of CO_2, 33, 34
  through sieve tubes (see also rate of translocation), 53, 58, 59
Fraxinus americana, 54, 59, 64
freezing and specimen preparation, 13, 14, 15
fructosans, 44, 72, 141
fruit, movement into, 58, 59, 118, 119, 134, 136, 137, 150–5

gibberellins, 121, 156
Gleichenia, 19
glutamine, 133
Glycine max, 14, 47, 51, 52, 155
Gossypium, 25, 58, 61, 62
grain growth, 146–50
grasses, transport in, 141
gravity, 78, 95
growth substances, effects on translocation, 120, 121
guttation, 103
gymnosperms, 9, 11, 19

half-width of tracer pulse, 42, 47, 48, 49
Helianthus, 47, 68
herbicide, transport of, 160, 161
Hordeum vulgare, 90, 92, 93, 94, 117, 132, 146, 148
hydraulic conductivity, 63, 74, 78
hydroids, 8, 19, 20
hyphae, trumpet hyphae, 2–5

import into leaves
  of carbohydrates, 125–30
  of mineral nutrients, 130–7
incipient plasmolysis, 78
indoleacetic acid (IAA), 120, 121
interfacial flow, 70
invertase, 72, 73, 136, 137, 142
ion movement
  into cells, 79–82
  into fruit, 155, 156
  in leaves, 102, 103, 104, 130, 131, 132
  in phloem, 104, 105, 106, 131, 132, 144, 145, 146
  into roots, 83–95
  into shoots, 92, 93
  into tubers, 144, 145, 146
  into xylem, 82–94

out of xylem, 102, 103, 132–7, 155, 156
  through xylem, 95–102
irradiance, 115, 127, 154
iron, 93
Isoetes, 8

kinins, 121
Knightia excelsa, 27

Laminaria, 2, 3, 5
lead chedate, 102
leaf
  analysis, 132
  growth, 123–38
  primordia, 122–5, 127, 128, 130
  sheath, 123, 125
legume fruit, 134, 136, 137, 155, 156, 157, 158
Lemna minor, 17, 18
leptoids, 7, 8
lignin, 19
liverworts, 18
loading of phloem, see phloem
Lolium, 114, 115, 127, 128, 129, 134, 136
Lupinus albus, 134, 136, 137, 140, 155, 156
Lycopersicon esculentum, 40, 41, 44, 45, 46, 111, 116, 118, 119, 127, 141, 150–4
Lycopodium, 8, 9

Macrocystis pyrifera, 4
magnesium, 132, 141, 157, 158, 159
malate, 35, 40
Malus sylvestris, 54, 99, 100, 101
mannitol, 54, 63, 64
margo, 22, 23
marrow, see *Cucurbita pepo*
mathematical models, 73, 74, 141, 142, 147
matric potential, 78, 79
maize, see *Zea mays*
micronutrients, 104
mineral nutrients in seeds, 157, 158, 159
mitochondria, 4, 5, 6, 8, 11
monosaccharides, 55
mosses, 8, 18, 19
Münch, 61

nacreous wall, 8, 9
Nernst equation, 82
Nicotiana tabacum, 35, 122, 123
nitrate, 56, 57, 95, 112, 113, 132–5
nitrate reductase, 132–5
nitrogen, 111, 112, 116, 132, 136, 137, 150, 156–9

fixation, 134, 136, 137
storage, 134, 138, 140, 141, 150
transport, 132–7, 150
nodal plexus, 125, 130
Nothofagus fusca, 21
nucleus, 4, 8, 11
nutrient deficiency, 83, 98
Nyphoides peltata, 15

Oats, see *Avena sativa*
Ophioglossum, 19
Oryza, 150
osmotic potential, 38, 78, 79, 97, 103, 110
oxidative phosphorylation, 104

pH, 33, 39, 54, 56, 57
p-protein, 8, 10, 11, 18, 65, 66, 70
palisade mesophyll, 42, 43
palms, 8, 54
partitioning of assimilates, 142, 164
Panicum maximum, 128, 129
peristaltic flow, 69
pesticides, 106, 160, 161
Phaseolus vulgaris, 100
phloem
  development, 2–18, 125, 129, 130
  entry of carbohydrates, 35–46, 63, 113, 114
  entry of ions, 37, 104
  evolution, 2–18
  kinetics of carbohydrate movement, 47–60
  mechanisms of movement, 60–71
  movement of ions, 104, 105, 106
  movement of nitrogen, 132–7, 155, 156
  movement out of, 72–3
  parencheyma, 8, 11, 15, 30
phosphoenolpyruvate (PEP) carboxylase, 35, 155
phosphorus, 37, 56, 57, 90–4, 97, 98, 99, 102, 106, 116, 130, 131, 132, 145, 146, 150, 151, 157, 158, 159
photorespiration, 35
photosynthesis, 33, 34, 35, 109, 110, 112, 125, 128, 129, 131, 143, 148, 150, 151, 154, 155
  in cereals, 125, 128, 131, 133, 134, 148, 150
  and rate of translocation, 45, 46, 109, 110, 112, 129, 143, 151, 154
photosynthetic phosphorylation, 82, 104
Picea, 96
Pinus, 22, 24
Pisum sativum, 73, 115, 116, 132, 133, 150, 155, 157, 158

Index

pits, 19, 21, 23, 27, 95
  bordered, 19, 21, 22, 23, 24, 95
plasmamembrane (plasmalemma), 2, 15,
  18, 28, 29, 35, 36, 63, 78, 80, 90
plasmodesma, 8, 9, 10, 11, 19, 28, 29, 30,
  35, 36, 40, 90
plastids, 4, 5, 11
Poiseville equation, 49
polysaccharides, 109
Polytrichium commune, 7, 20
poplar, see *Populus nigra*
Populus nigra, 97, 98, 99, 102
potassium, 39, 54, 83, 84, 85, 86, 93, 94,
  97, 100, 104, 111, 112, 116, 130, 141,
  157, 158, 159
potato, see *Solanum tuberosum*
pressure
  flow, 61, 63–7, 110
  in sieve cells, 63–7
  in xylem, 90, 95–102
  potential, 78
primary cell wall, 19
proton cotransport, 39
protophloem, 123
protostele, 19
protoxylem, 19, 23, 25
Pseudotsuga, 21
Pteridium, 19

Quercus rubra, 54

raffinase, 54
rate of translocation, 45, 53, 58–64, 127,
  128, 129, 144, 150–5
resistance
  to flow, 25, 38, 54, 90, 144
  to transport, 34, 87, 89, 97, 102, 147
respiration, 59, 67, 82, 117, 118, 119,
  144, 150, 154, 155
Rhizobium, 134
rhizoids, 18
ribosomes, 8, 11
ribulose bisphosphate (RUBP)
  carboxylase, 34, 35, 40
rice, see *Oryza*
Ricinus communis, 54, 55, 56, 84, 85, 86
ring porous species, 20
ringing of phloem, 1
Robinia pseudoacacia, 55
roots, 18, 59
  exudate, 142
  fibrous, 142
  growth of, 95, 141, 142
  hairs, 18
  ion uptake by, 83–95, 111, 112, 113,

116, 117, 132–7
  nodules, 134, 136, 137
  pressure, 83, 95
  storage, 142
root:shoot ratio, 142
rubidium, 104

Saccharum officinarum, 72, 73, 139
Salix viminalis, 64, 120
salt glands, 104
secondary
  phloem, 20
  xylem, 20, 21
seeds, transport from, 155–9
Selaginella, 8
Senecio vulgaris, 30
senescence, 130, 132, 150, 155
short wave radiation, 111, 112
sieve
  areas, 6
  cells (sieve elements, sieve tubes),
    6–18, 28, 30, 35, 37, 41, 42, 47–57,
    60–73, 104, 106, 125, 139, 161, 163
  plates, 4, 5, 6, 12, 14
  pores, 6, 14, 15, 16, 54, 65, 67
sinks, 61, 63, 72, 117, 118, 120, 121, 143,
  144, 151, 163
slime, 10, 11, 18
sodium, 157, 158
Solanum tuberosum, 58, 73, 87, 105, 109,
  139
sorbitol, 54
Sorghum sudanense, 115
sources, 61, 72, 147, 151, 163
specimen preparation, problems of, 2, 11,
  12, 13, 16, 18
speed of movement through phloem, 4,
  13, 48–53, 60, 61, 63, 64, 65, 112–15,
  127
speed of movement through xylem, 98,
  99, 102
spongy mesophyll, 42, 43
stachyose, 54, 64
starch, 11, 34, 35, 44, 72, 73, 112, 133,
  135, 141, 143
stems, 47, 87, 96–102, 123, 124, 133, 135,
  136, 139, 140, 141, 149, 150, 151
stolons, 144
stomata, 18, 88, 112, 117
storage
  of carbohydrates, 47, 48, 72, 73, 139,
    141, 142
  of mineral nutrients, 47, 134, 136, 138,
    139, 141, 142
  organs, 72, 73, 142–59

168

strontium, 104, 105, 145
suberin, 87, 89
sucrose, 34–7, 39, 54–7, 59–62, 64, 65, 72, 73, 113, 114, 119, 131, 133, 135, 150, 152, 153
sugar beet, see *Beta vulgaris*
sugar cane, see *Saccharum officinarum*
sulphate, 104, 132
symplasm (symplast), 28, 35–8, 72, 73, 81, 89, 90, 103

temperature
  effect of movement into fruit, 118, 119
  effect on respiration, 67, 117, 119
  effect on translocation, 67, 68, 118, 119, 120, 152, 153
  effect on xylem movement, 95, 117
  roots, 116, 117
  water potential, 117
thickening of xylem, 24, 25
thylakoid membrane, 11
tillers on cereals and grasses, 139, 140, 141
tomato, see *Lycopersicon esculentum*
tonoplast, 14, 15, 17, 18, 90
torus, 22, 23, 24, 95
tracheid, 2, 9, 19, 20, 21
transcellular strands, 13, 18, 67, 69, 70
transfer cells, 29, 30, 35, 36, 146, 147, 148
transpiration, 37, 38, 87, 88, 90, 95, 97, 102, 111, 112, 144, 145, 155, 156
transport systems, effect
  on growth of cereal grains, 146–50
  on growth of fruit, 150–5
  on growth of roots, 141, 142
  on growth of stem apices and leaves, 123–41
  on growth of stems, 140, 142
  on growth of storage organs and fruit, 142–55
  of growth substances, 112, 120, 121
  on growth of tubers, 143–6

on ion uptake, 83–6, 91–5, 111, 112, 113
of light, 112–16
of temperature, 67, 68, 117–20, 152, 153
of water, 109, 110, 112
Triticum aestivum, 35, 38, 48, 49, 106, 110, 122, 123, 124, 126, 132, 139, 140, 146–51
tuber growth, 73, 143–6
turgid cells, 78
turgor pressure, 2, 63, 65, 78, 79, 95
tyloses, 24

Ulmus, 27

vascular bundles (veins) in leaf, 36, 37, 38, 42, 43, 124, 125, 129
vessel, 2, 25, 26, 27
Vicia faba, 35, 42, 43
viscosity, 49, 56, 64, 67

water
  content of cells, 76–9
  deficit, 87, 109, 110, 143, 144
  free energy of, 76, 77
  movement into cells, 76–79
  movement into xylem, 82, 83, 89, 90
  movement through xylem, 87, 88, 90, 95–8, 102, 110, 155, 156
  potential, 38, 57, 63, 77, 78, 79, 82, 83, 87, 90, 96, 97, 109, 110, 117

Xanthium pennsylvanicum, 134, 135
xylem, 1, 2, 18–27
  entry of water and ions, 82–90
  transport through, 95–102

Yucca flaccida, 54, 57

Zea mays, 42, 52, 53, 73, 113, 115, 146, 150